Student Solutions Manual

for use with

Elementary Statistics
A Step by Step Approach

Fifth Edition

Allan G. Bluman

Community College of Allegheny County

Prepared by

Sally H. Robinson

South Plains College

 **Higher Education**

Boston   Burr Ridge, IL   Dubuque, IA   Madison, WI   New York   San Francisco   St. Louis
Bangkok   Bogotá   Caracas   Kuala Lumpur   Lisbon   London   Madrid   Mexico City
Milan   Montreal   New Delhi   Santiago   Seoul   Singapore   Sydney   Taipei   Toronto

The **McGraw·Hill** Companies

Student Solutions Manual for use with
ELEMENTARY STATISTICS: A STEP BY STEP APPROACH, FIFTH EDITION
ALLAN G. BLUMAN

Published by McGraw-Hill Higher Education, an imprint of The McGraw-Hill Companies, Inc.,
1221 Avenue of the Americas, New York, NY 10020. Copyright © The McGraw-Hill Companies,
Inc., 2004. All rights reserved.

1 2 3 4 5 6 7 8 9 0 QPD/QPD 0 9 8 7 6 5 4 3

ISBN 0-07-254912-2

www.mhhe.com

Preface

This Student Solutions Manual provides solutions to **selected even and all odd exercises** plus answers to all quiz questions in *Elementary Statistics: A Step by Step Approach, 5e* by Allan G. Bluman. Solutions are worked out step by step where appropriate and generally follow the same procedures used in the examples in the textbook. Answers may be carried to several decimal places to increase accuracy and to facilitate checking. See your instructor for specific rounding rules. Graphs are included with the solutions when appropriate or required. They are intended to convey a general idea and may not be to scale.

Caution: Answers generated using graphing calculators such as the TI-83 may vary from those shown in this manual.

To maximize the assistance provided in this manual, you should:

1. Read each section of the text carefully, noting examples and formulas.

2. Begin working the exercises using textbook examples and class notes as your guide, then refer to the answers in this manual.

3. Many instructors require students to interpret their answers within the context of the problem. These interpretations are powerful tools to understanding the meaning and purpose of each calculation. You should attempt to interpret each calculation even if you are not required to do so.

4. Be sure to show your work. When checking your work for errors, you will need to review each step. When preparing for exams, reviewing each step helps you to recall the process involved in producing each calculation.

5. As you gain confidence and understanding, you should attempt to work exercises without referring to examples or notes. Check each answer in the solutions manual before beginning the next exercise.

6. Slight variations between your answers and the answers in this manual are probably due to rounding differences and should not be a cause for concern. If you are concerned about these variations, check each step of your calculation.

7. Many errors can be traced to the improper application of the rules for order of operations. You should first attempt to determine where and how your error occurred, because diagnosing your error will increase your understanding and prevent future errors. See your instructor if you are unsure of the location or cause of your error.

Also available for student purchase to improve your success in this course when using *Elementary Statistics: A Step by Step Approach, 5e* by Allan G. Bluman:

Student Study Guide

By Pat Foard of South Plains College, this study guide will assist students in understanding and reviewing key concepts and preparing for exams. It emphasizes all important concepts contained in each chapter, includes explanations, and provides opportunities for students to test their understanding by completing related exercises and problems.

Critical Thinking Workbook

By James Condor of Manatee Community College, this workbook provides a number of additional challenging problems for students to solve that are drawn from real-world applications. Problems are tied to each chapter of the textbook and highlight and reinforce key concepts.

MINITAB Manual

By Gerry Moultine of Northwood University, this manual provides the student with how-to information on data and file management, conducting various statistical analyses, and creating presentation-style graphics while following each text chapter.

TI-83 and TI-83 Plus Graphing Calculator Manual

By Carolyn Meitler of Concordia University Wisconsin, this patient, practical manual teaches students to learn about statistics and solve problems using these calculators while following each text chapter.

Excel Manual for Office 2000

By Renee Goffinet and Virginia Koehler of Spokane Falls Community College, this workbook is specially designed to accompany the textbook and provides additional practice in applying the chapter concepts while using Excel.

Videos

New to this edition are text-specific videos available on VHS and CD-ROM that demonstrate key concepts and worked-out exercises from the text plus tutorials in using the TI-83 Plus Calculator, Excel, and MINITAB, in a dynamic, engaging format.

Sally H. Robinson

Contents

Solutions to the Exercises

REVIEW EXERCISES - CHAPTER 1

1. Descriptive statistics describes a set of data. Inferential statistics uses a set of data to make predictions about a population.

3. Answers will vary.

5. When the population is large, the researcher saves time and money using samples. Samples are used when the units must be destroyed.

6.
 a. inferential   e. inferential
 b. descriptive   f. inferential
 c. descriptive   g. descriptive
 d. descriptive   h. inferential

7.
 a. ratio   f. nominal
 b. ordinal   g. ratio
 c. interval   h. ratio
 d. ratio   i. ordinal
 e. ratio   j. ratio

8.
 a. qualitative   e. quantitative
 b. quantitative   f. quantitative
 c. qualitative   g. quantitative
 d. quantitative

9.
 a. discrete   e. continuous
 b. continuous   f. continuous
 c. discrete   g. discrete
 d. continuous

11. Random samples are selected by using chance methods or random numbers. Systematic samples are selected by numbering each subject and selecting every kth number. Stratified samples are selected by dividing the population into groups and selecting from each group. Cluster samples are selected by using intact groups called clusters.

12.
 a. cluster   d. systematic
 b. systematic e. stratified
 c. random

13. Answers will vary.

15. Answers will vary.

17.
 a. experimental   c. observational
 b. observational   d. experimental

19. Answers will vary. Possible answers include:
(a) overall health of participants, amount of exposure to infected individuals through the workplace or home
(b) gender and/or age of driver, time of day
(c) diet, general health, heredity factors
(d) amount of exercise, heredity factors

21. Claims can be proven only if the entire population is used.

23. Since the results are not typical, the advertisers selected only a few people for whom the product worked extremely well.

25. "74% more calories" than what? No comparison group is stated.

27. What is meant by "24 hours of acid control"?

29. Possible reasons for conflicting results: The amount of caffeine in the coffee or tea or the brewing method.

31. Answers will vary.

CHAPTER 1 QUIZ
1. True
2. False, it is a data value.
3. False, the highest level is ratio.
4. False, it is stratified sampling.
5. False, it is a quantitative variable.
6. True
7. False, it is 5.5-6.5 inches.
8. c.
9. b.
10. d.
11. a.
12. c.
13. a.
14. descriptive, inferential
15. gambling, insurance
16. population
17. sample

18.
a. saves time
b. saves money
c. use when population is infinite

19.
a. random   c. cluster
b. systematic   d. stratified

20. quasi-experimental

21. random

22.
a. inferential   d. descriptive
b. descriptive   e. inferential
c. inferential

23.
a. ratio   d. ratio
b. ordinal   e. nominal
c. interval

24.
a. continuous   d. continuous
b. discrete   e. continuous
c. discrete   f. discrete

25.
a. 3.15-3.25   d. 0.265-0.275
b. 17.5-18.5   e. 35.5-36.5
c. 8.5-9.5

EXERCISE SET 2-2

1. Frequency distributions are used to:
    1. organize data in a meaningful way
    2. determine the shape of the distribution
    3. facilitate computation procedures for finding descriptive measures such as the mean
    4. draw charts and graphs
    5. make comparisons between data sets

3.
a. $10.5 - 15.5$, $\frac{11+15}{2} = \frac{26}{2} = 13$, $15.5 - 10.5 = 5$
b. $16.5 - 39.5$, $\frac{17+39}{2} = \frac{56}{2} = 28$, $39.5 - 16.5 = 23$
c. $292.5 - 353.5$, $\frac{292+353}{2} = \frac{646}{2} = 323$, $353.5 - 292.5 = 61$
d. $11.75 - 14.75$, $\frac{11.75+14.75}{2} = \frac{26.5}{2} = 13.25$, $14.75 - 11.75 = 3$
e. $3.125 - 3.935$, $\frac{3.13+3.93}{2} = \frac{7.06}{2} = 3.53$, $3.935 - 3.125 = 0.81$

5.
a. Class width is not uniform.
b. Class limits overlap, and class width is not uniform.
c. A class has been omitted.
d. Class width is not uniform.

7.

Class	Tally	f	Percent
W	⦀⦀⦀ |	16	32%
BL	⦀⦀ |||	13	26%
BR	⦀ ||||	9	18%
Y	⦀ |	6	12%
G	⦀ |	6	12%
		50	100%

9.

Class	Boundaries	f	cf
0	- 0.5 - 0.5	5	5
1	0.5 - 1.5	8	13
2	1.5 - 2.5	10	23
3	2.5 - 3.5	2	25
4	3.5 - 4.5	3	28
5	4.5 - 5.5	2	30
		30	

11. $H = 780$ $L = 746$
Range $= 780 - 746 = 34$
Width $= 34 \div 6 = 5.\overline{6}$ or 6

11. continued

Limits	Boundaries	f	cf
746 - 751	745.5 - 751.5	4	4
752 - 757	751.5 - 757.5	4	8
758 - 763	757.5 - 763.5	7	15
764 - 769	763.5 - 769.5	6	21
770 - 775	769.5 - 775.5	6	27
776 - 781	775.5 - 781.5	3	30
		30	

13. $H = 70$ $L = 27$
Range $= 70 - 27 = 43$
Width $= 43 \div 7 = 6.1$ or 7

Limits	Boundaries	f	cf
27 - 33	26.5 - 33.5	7	7
34 - 40	33.5 - 40.5	14	21
41 - 47	40.5 - 47.5	15	36
48 - 54	47.5 - 54.5	11	47
55 - 61	54.5 - 61.5	3	50
62 - 68	61.5 - 68.5	3	53
69 - 75	68.5 - 75.5	2	55
		55	

15.

Limits	Boundaries	f	cf
0 - 19	-0.5 - 19.5	13	13
20 - 39	19.5 - 39.5	18	31
40 - 59	39.5 - 59.5	10	41
60 - 79	59.5 - 79.5	5	46
80 - 99	79.5 - 99.5	3	49
100 - 119	99.5 - 119.5	1	50
		50	

17. $H = 11,413$ $L = 150$
Range $= 11,413 - 150 = 11,263$
Width $= 11,263 \div 10 = 1126.3$ or 1127

Limits	Boundaries	f	cf
150 - 1276	149.5 - 1276.5	2	2
1277 - 2403	1276.5 - 2403.5	2	4
2404 - 3530	2403.5 - 3530.5	5	9
3531 - 4657	3530.5 - 4657.5	8	17
4658 - 5784	4657.5 - 5784.5	7	24
5785 - 6911	5784.5 - 6911.5	3	27
6912 - 8038	6911.5 - 8038.5	7	34
8039 - 9165	8038.5 - 9165.5	3	37
9166 - 10,292	9165.5 - 10,292.5	3	40
10,293 - 11,419	10,292.5 - 11,419.5	2	42
		42	

EXERCISE SET 2-3

1.

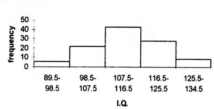

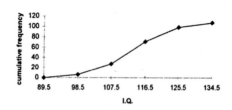

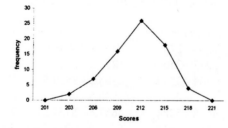

Eighty applicants do not need to enroll in the summer programs.

3.

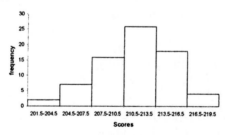

3. continued

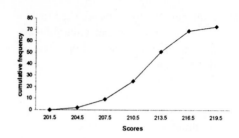

The distribution appears to be slightly left skewed.

5.

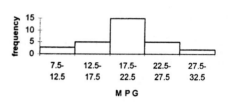

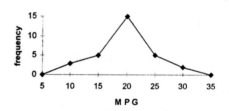

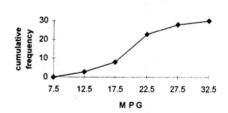

7.

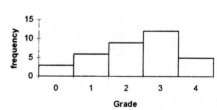

7. continued

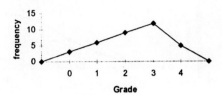

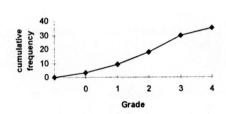

Yes, 26 out of the 35 students can enroll in the next course.

9.

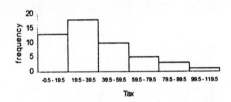

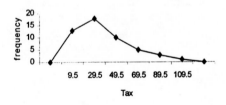

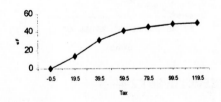

The majority of the states charge less than 40 cents per pack.

11.

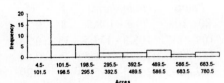

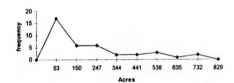

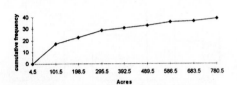

The majority of the parks had between 4.5 and 101.5 thousand acres.

13.

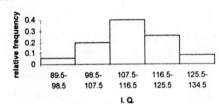

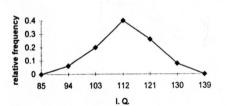

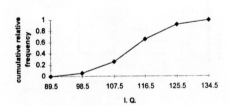

The proportion of applicants who need to enroll in a summer program is 0.26 or 26%.

15. H = 270   L = 80
Range = 270 − 80 = 190
Width = 190 ÷ 7 = 27.1 or 28
Use width = 29 (rule 2)

15. continued

Limits	Boundaries	f	rf	crf
80 - 108	79.5 - 108.5	8	0.17	0.17
109 - 137	108.5 - 137.5	13	0.28	0.45
138 - 166	137.5 - 166.5	2	0.04	0.49
167 - 195	166.5 - 195.5	9	0.20	0.69
196 - 224	195.5 - 224.5	10	0.22	0.91
225 - 253	224.5 - 253.5	2	0.04	0.95
254 - 282	253.5 - 282.5	2	0.04	0.99*
			0.99*	

*due to rounding

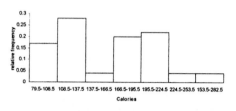

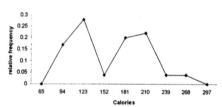

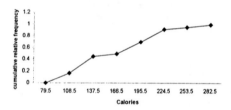

17.

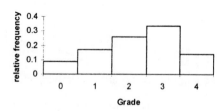

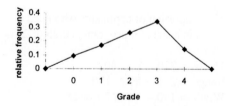

17. continued

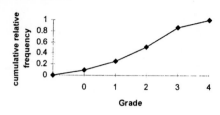

The proportion of students who cannot meet the requirement for the next course is 0.26 or 26%.

19.

Limits	Boundaries	X_m	f	cf
22 - 24	21.5 - 24.5	23	1	1
25 - 27	24.5 - 27.5	26	3	4
28 - 30	27.5 - 30.5	29	0	4
31 - 33	30.5 - 33.5	32	6	10
34 - 36	33.5 - 36.5	35	5	15
37 - 39	36.5 - 39.5	38	3	18
40 - 42	39.5 - 42.5	41	2	20
			20	

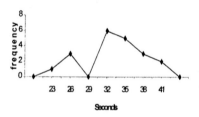

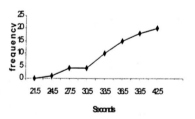

EXERCISE SET 2-4

1.

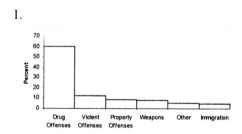

6

3.

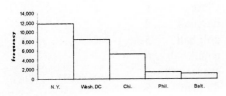

5.

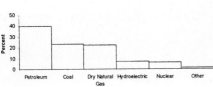

7.

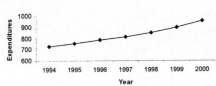

9.

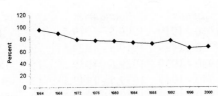

The graph shows a decline in the percents of registered voters voting in Presidential elections.

11.

Personal Residence	7.8%	28.08°
Liquid Assets	5.0%	18.0°
Pension Accounts	6.9%	24.84°
Stocks, Funds, and Trusts	31.6%	113.76°
Business & Real Estate	46.9%	168.84°
Miscellaneous	1.8%	6.48°
	100.0%	360.00°

11. continued

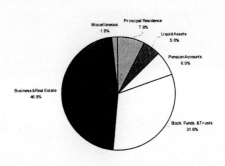

13.

Career change	34%	122.4°
New job	29%	104.4°
Start business	21%	75.6°
Retire	16%	57.6°
	100%	360.0°

Pie chart:

Pareto chart:

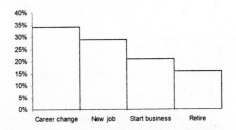

The Pareto chart is better at showing a comparison between categories.

15.

```
4 | 2 3
4 | 6 6 7 8 9 9
5 | 0 1 1 1 1 2 2 4 4 4 4 4
5 | 5 5 5 5 6 6 6 7 7 7 7 8
6 | 0 1 1 1 2 4 4
6 | 5 8 9
```

15. continued
The majority of the Presidents were in their 50's at inauguration.

17.

Variety 1												Variety 2				
								2	1	3	8					
							3	0	2	5						
			9	8	8	5	2	3	6	8						
					3	3	1	4	1	2	5	5				
9	9	8	5	3	3	2	1	0	5	0	3	5	5	6	7	9
								6	2	2						

The distributions are similar but variety 2 seems to be more variable than variety 1.

19.

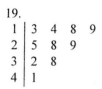

1	3	4	8	9
2	5	8	9	
3	2	8		
4	1			

21.

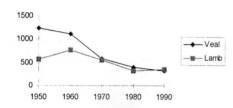

In 1950, veal production was considerably higher than lamb. By 1970, production was approximately the same for both.

23.

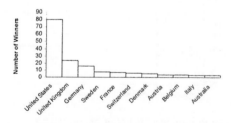

REVIEW EXERCISES - CHAPTER TWO

1.

Class	f
Newspaper	7
Television	5
Radio	7
Magazine	6
	25

3.

Class	f
baseball	4
golf ball	5
tennis ball	6
soccer ball	5
football	5
	25

5.

Class	f	cf
11	1	1
12	2	3
13	2	5
14	2	7
15	1	8
16	2	10
17	4	14
18	2	16
19	2	18
20	1	19
21	0	19
22	1	20
	20	

7.

Limits	Boundaries	f	cf
1910 - 1919	1909.5 - 1919.5	1	1
1920 - 1929	1919.5 - 1929.5	2	3
1930 - 1939	1929.5 - 1939.5	15	18
1940 - 1949	1939.5 - 1949.5	12	30
1950 - 1959	1949.5 - 1959.5	20	50
1960 - 1969	1959.5 - 1969.5	18	68
1970 - 1979	1969.5 - 1979.5	18	86
1980 - 1989	1979.5 - 1989.5	6	92
1990 - 1999	1989.5 - 1999.5	8	100
		100	

9.

Limits	Boundaries	f	cf
170 - 188	169.5 - 188.5	11	11
189 - 207	188.5 - 207.5	9	20
208 - 226	207.5 - 226.5	4	24
227 - 245	226.5 - 245.5	5	29
246 - 264	245.5 - 264.5	0	29
265 - 283	264.5 - 283.5	0	29
284 - 302	283.5 - 302.5	0	29
303 - 321	302.5 - 321.5	1	30
		30	

11.

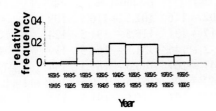

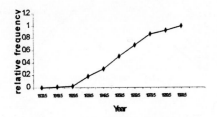

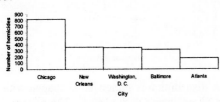

13.

15.

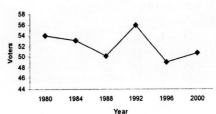

The minimum wage has increased over the years with the largest increase occurring between 1975 and 1980.

17.

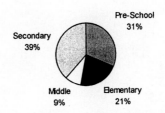

The highest percent of voting-age population that voted in a Presidential election occurred in 1992, while the lowest percent occurred in 1996.

19.

The fewest number of students were enrolled in the middle school field, and more students were in the secondary field than any other field.

21.

```
1 | 2  4
1 | 6  7  8  8  9
2 | 0  2  3  4
2 | 5  5  5  6  6  9  9
3 | 2  3
3 | 5  7  8  8  9
```

CHAPTER 2 QUIZ

1. False
2. False
3. False
4. True
5. True
6. False
7. False
8. c.
9. c.
10. b.
11. b.
12. Categorical, ungrouped, grouped
13. 5, 20
14. categorical
15. time series

16. stem and leaf plot
17. vertical or y
18.

	f	cf
H	6	6
A	5	11
M	6	17
C	8	25
	25	

19.

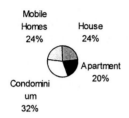

Mobile Homes 24%
House 24%
Apartment 20%
Condominium 32%

20.

Class	f	cf
0.5 − 1.5	1	1
1.5 − 2.5	5	6
2.5 − 3.5	3	9
3.5 − 4.5	4	13
4.5 − 5.5	2	15
5.5 − 6.5	6	21
6.5 − 7.5	2	23
7.5 − 8.5	3	26
8.5 − 9.5	4	30
	30	

21.

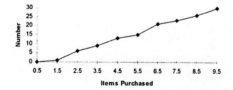

22.

Class	Boundaries	mp	f	cf
102 − 116	101.5 − 116.5	109	4	4
117 − 131	116.5 − 131.5	124	3	7
132 − 146	131.5 − 146.5	139	1	8
147 − 161	146.5 − 161.5	154	4	12
162 − 176	161.5 − 176.5	169	11	23
177 − 191	176.5 − 191.5	184	7	30
			30	

23.

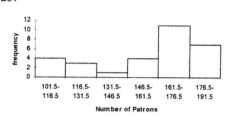

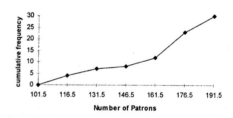

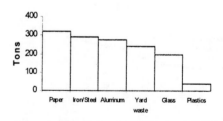

The distribution is somewhat U-shaped with a peak occurring in the 161.5 - 176.5 class.

24.

Paper, Iron/Steel, Aluminum, Yard waste, Glass, Plastics

10

25.

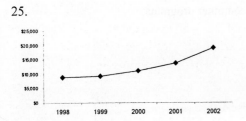

26.

```
1 | 5  9
2 | 6  8
3 | 1  5  8  8  9
4 | 1  7  8
5 | 3  3  4
6 | 2  3  7  8
7 | 6  9
8 | 6  8  9
9 | 8
```

Note: Answers may vary slightly due to rounding, TI 83's, or computer programs.

EXERCISE SET 3-2

1.
$$\overline{X} = \frac{\Sigma X}{n} = \frac{93.09}{25} = 3.7236 \approx 3.724$$

MD: 3.57, 3.64, 3.64, 3.65, 3.66, 3.67, 3.67, 3.68, 3.7, 3.7, 3.7, 3.73, **3.73**, 3.74, 3.74, 3.74, 3.75, 3.76, 3.77, 3.78, 3.78, 3.8, 3.8, 3.83, 3.86

Mode: 3.7 and 3.74   MR: $\frac{3.57 + 3.86}{2} = 3.715$

3.
$$\overline{X} = \frac{\Sigma X}{n} = \frac{136}{9} = 15.1$$

MD: 1, 2, 3, 3, **7**, 11, 18, 30, 61

Mode = 3   MR $= \frac{1+61}{2} = 31$

The median is probably the best measure of average because 61 is an extremely large data value and makes the mean artificially high.

5.
$$\overline{X} = \frac{\Sigma X}{n} = \frac{218}{16} = 13.625$$

MD: 1, 5, 6, 7, 8, 10, 10, **11, 12**, 14, 15, 16, 18, 22, 28, 35

MD $= \frac{11 + 12}{2} = 11.5$

Mode: 10   MR $= \frac{1 + 35}{2} = 18$

7.
$$\overline{X} = \frac{\Sigma X}{n} = \frac{79.6}{12} = 6.63$$

MD: 5.4, 5.4, 6.2, 6.2, 6.4, **6.4, 6.5**, 7.0, 7.2, 7.2, 7.7, 8.0

MD $= \frac{6.4 + 6.5}{2} = 6.45$

Mode: no mode   MR $= \frac{5.4 + 8.0}{2} = 6.7$

9.
$$\overline{X} = \frac{\Sigma X}{n} = \frac{238,512}{42} = 5678.9$$

MD: 150, 885, ..., **5315, 5370**, ..., 11070, 11413

MD $= \frac{5315 + 5370}{2} = 5342.5$

Mode: 4450   MR $= \frac{150 + 11,413}{2} = 5781.5$

The distribution is skewed to the right.

11.
For Year 1:

$$\overline{X} = \frac{\Sigma X}{n} = \frac{24,911}{27} = 922.6$$

MD = 527

Mode: no mode $MR = \frac{69+4192}{2} = 2130.5$

For Year 2:

$$\overline{X} = \frac{\Sigma X}{n} = \frac{24,615}{2} = 911.7$$

MD = 485

Mode: 1430 $MR = \frac{70+4040}{2} = 2055$

The mean, median, and midrange of the traffic fatalities for Year 2 are somewhat less than those for the Year 1 fatalities, indicating that the number of fatalities has decreased.

13.

Class Limits	Boundaries	X_m	f	$f \cdot X_m$
202 - 204	201.5 - 204.5	203	2	406
205 - 207	204.5 - 207.5	206	7	1442
208 - 210	207.5 - 210.5	209	16	3344
211 - 213	210.5 - 213.5	212	26	5512
214 - 216	213.5 - 216.5	215	18	3870
217 - 219	216.5 - 219.5	218	4	872
			73	15,446

$$\overline{X} = \frac{\sum f \cdot X_m}{n} = \frac{15,446}{73} = 211.6$$

modal class: $211 - 213$

15.

Class	Boundaries	X_m	f	$f \cdot X_m$
0 - 2	-0.5 - 2.5	1	2	2
3 - 5	2.5 - 5.5	4	6	24
6 - 8	5.5 - 8.5	7	12	84
9 - 11	8.5 - 11.5	10	5	50
12 - 14	11.5 - 14.5	13	3	39
			28	199

$$\overline{X} = \frac{\sum f \cdot X_m}{n} = \frac{199}{28} = 7.1$$

modal class: $5.5 - 8.5$

17.

Boundaries	X_m	f	$f \cdot X_m$
52.5 – 63.5	58	6	348
63.5 – 74.5	69	12	828
74.5 – 85.5	80	25	2000
85.5 – 96.5	91	18	1638
96.5 – 107.5	102	14	1428
107.5 – 118.5	113	5	565
		80	6807

$$\overline{X} = \frac{\sum f \cdot X_m}{n} = \frac{6807}{80} = 85.1$$

modal class: $74.5 - 85.5$

19.

Class Limits	Boundaries	X_m	f	$f \cdot X_m$
13 – 19	12.5 – 19.5	16	2	32
20 – 26	19.5 – 26.5	23	7	161
27 – 33	26.5 – 33.5	30	12	360
34 – 40	33.5 – 40.5	37	5	185
41 – 47	40.5 – 47.5	44	6	264
48 – 54	47.5 – 54.5	51	1	51
55 – 61	54.5 – 61.5	58	0	0
62 – 68	61.5 – 68.5	65	2	130
			35	1183

$$\overline{X} = \frac{\sum f \cdot X_m}{n} = \frac{1183}{35} = 33.8$$

modal class: $26.5 - 33.5$

21.

Boundaries	X_m	f	$f \cdot X_m$
15.5 – 18.5	17	14	238
18.5 – 21.5	20	12	240
21.5 – 24.5	23	18	414
24.5 – 27.5	26	10	260
27.5 – 30.5	29	15	435
30.5 – 33.5	32	6	192
		75	1779

$$\overline{X} = \frac{\sum f \cdot X_m}{n} = \frac{1779}{75} = 23.7$$

modal class: $21.5 - 24.5$

23.

Limits	Boundaries	X_m	f	$f \cdot X_m$
27 - 33	26.5 - 33.5	30	7	210
34 - 40	33.5 - 40.5	37	14	518
41 - 47	40.5 - 47.5	44	15	660
48 - 54	47.5 - 54.5	51	11	561
55 - 61	54.5 - 61.5	58	3	174
62 - 68	61.5 - 68.5	65	3	195
69 - 75	68.5 - 75.5	72	2	144
			55	2462

$$\overline{X} = \frac{\sum f \cdot X_m}{n} = \frac{2462}{55} = 44.8$$

modal class: $40.5 - 47.5$

25.

Limits	Boundaries	X_m	f	$f \cdot X_m$
0 - 19	-0.5 - 19.5	9.5	13	123.5
20 - 39	19.5 - 39.5	29.5	18	531.0
40 - 59	39.5 - 59.5	49.5	10	495.0
60 - 79	59.5 - 79.5	69.5	5	347.5
80 - 99	79.5 - 99.5	89.5	3	268.5
100 - 119	99.5 - 119.5	109.5	1	109.5
			50	1875.0

$$\overline{X} = \frac{\sum f \cdot X_m}{n} = \frac{1875}{50} = 37.5$$

modal class: $19.5 - 39.5$

27.
$$\overline{X} = \frac{\sum w \cdot X}{\sum w} = \frac{3(3.33) + 3(3.00) + 2(2.5) + 2.5(4.4) + 4(1.75)}{3 + 3 + 2 + 2.5 + 4} = \frac{41.99}{14.5} = 2.896$$

29.
$$\overline{X} = \frac{\sum w \cdot X}{\sum w} = \frac{9(427000) + 6(365000) + 12(725000)}{9 + 6 + 12} = \frac{14,733,000}{27} = \$545,666.67$$

31.
$$\overline{X} = \frac{\sum w \cdot X}{\sum w} = \frac{1(62) + 1(83) + 1(97) + 1(90) + 2(82)}{6} = \frac{496}{6} = 82.7$$

33.
a. Median
b. Mean
c. Mode
d. Mode
e. Mode
f. Mean

35.
Both could be true since one could be using the mean for the average salary, and the other could be using the mode for the average.

37.

$5 \cdot 8.2 = 41$

$6 + 10 + 7 + 12 + x = 41$

$x = 6$

39.

a. $\dfrac{2}{\frac{1}{30} + \frac{1}{45}} = 36$ mph

b. $\dfrac{2}{\frac{1}{40} + \frac{1}{25}} = 30.77$ mph

c. $\dfrac{2}{\frac{1}{50} + \frac{1}{10}} = \$16.67$

41.

$$\sqrt{\frac{8^2 + 6^2 + 3^2 + 5^2 + 4^2}{5}} = \sqrt{30} = 5.48$$

EXERCISE SET 3-3

1.

The square root of the variance is equal to the standard deviation.

3.

σ^2, σ

5.

When the sample size is less than 30, the formula for the true standard deviation of the sample will underestimate the population standard deviation.

7.

$R = 15 - 6 = 9$

$$s^2 = \frac{\sum X^2 - \frac{(\sum X)^2}{n}}{n-1} = \frac{1209 - \frac{(117)^2}{12}}{12-1} = \frac{68.25}{11} = 6.20$$

$s = \sqrt{6.20} = 2.5$

9.

For Temperature:

$R = 61 - 29 = 32$

$$s^2 = \frac{\sum X^2 - \frac{(\sum X)^2}{n}}{n-1} = \frac{20{,}777 - \frac{441^2}{10}}{10-1} = 147.66$$

$s = \sqrt{147.66} = 12.15$

For Precipitation:

$R = 5.1 - 1.1 = 4.0$

$$s^2 = \frac{\sum X^2 - \frac{(\sum X)^2}{n}}{n-1} = \frac{86.13 - \frac{26.3^2}{10}}{10-1} = 1.88$$

9. continued
$s = \sqrt{1.88} = 1.37$

Temperature is more variable.

11.
$R = 46 - 16 = 30$

$s^2 = \dfrac{\sum X^2 - \frac{(\sum X)^2}{n}}{n-1} = \dfrac{9677 - \frac{313^2}{11}}{11-1} = \dfrac{770.727}{10} = 77.1$

$s = \sqrt{77.1} = 8.8$

13.
$R = 22 - 1 = 21$

$s^2 = \dfrac{\sum X^2 - \frac{(\sum X)^2}{n}}{n-1} = \dfrac{1061 - \frac{89^2}{15}}{15-1} = 38.1$

$s = \sqrt{38.1} = 6.2$

15.
For l995:
$R = 4192 - 69 = 4123$

$s^2 = \dfrac{\sum X^2 - \frac{(\sum X)^2}{n}}{n-1} = \dfrac{49,784,885 - \frac{24,911^2}{27}}{27-1} = 1,030,817.63$

$s = \sqrt{1,030,817.63} = 1015.3$

For 1996:
$R = 4040 - 70 = 3970$

$s^2 = \dfrac{\sum X^2 - \frac{(\sum X)^2}{n}}{n-1} = \dfrac{48,956,875 - \frac{24,615^2}{27}}{27-1} = 1,019,853.85$

$s = \sqrt{1,019,853.85} = 1009.9$

The fatalities in 1995 are more variable.

17.
$R = 11,413 - 150 = 11,263$

$s^2 = \dfrac{\sum X^2 - \frac{(\sum X)^2}{n}}{n-1} = \dfrac{1,659,371,050 - \frac{238,512^2}{42}}{42-1} = \dfrac{304,895,475.1}{41} = 7,436,475.003$

$s = \sqrt{7,436,475.003} = 2726.99 \text{ or } 2727$

19.

X_m	f	$f \cdot X_m$	$f \cdot X_m^2$
16	2	32	512
23	7	161	3703
30	12	360	10800
37	5	185	6845
44	6	264	11616
51	1	51	2601
58	0	0	0
65	2	130	8450
	35	1183	44527

$$s^2 = \frac{\sum f \cdot X_m^2 - \frac{(\sum f \cdot X_m)^2}{n}}{n-1} = \frac{44{,}527 - \frac{1183^2}{35}}{35-1} = \frac{4541.6}{34} = 133.58$$

$$s = \sqrt{133.58} = 11.6$$

21.

Class	X_m	f	$f \cdot X_m$	$f \cdot X_m^2$
0 - 2	1	1	1	1
3 - 5	4	3	12	48
6 - 8	7	5	35	245
9 - 11	10	14	140	1400
12 - 14	13	6	78	1014
		29	266	2708

$$s^2 = \frac{\sum f \cdot X^2 - \frac{(\sum f \cdot X)^2}{n}}{n-1} = \frac{2708 - \frac{266^2}{29}}{29-1} = \frac{268.1379}{28} = 9.58$$

$$s = \sqrt{9.58} = 3.1$$

23.

X_m	f	$f \cdot X_m$	$f \cdot X_m^2$
58	6	348	20184
69	12	828	57132
80	25	2000	160000
91	18	1638	148058
102	14	1428	145656
112	5	565	63845
	80	6807	595875

$$s^2 = \frac{\sum f \cdot X_m^2 - \frac{(\sum f \cdot X_m)^2}{n}}{n-1} = \frac{595875 - \frac{6807^2}{80}}{80-1} = \frac{16684.39}{79} = 211.2$$

$$s = \sqrt{211.2} = 14.5$$

25.

X_m	f	$f \cdot X_m$	$f \cdot X_m^2$
56	2	112	6272
61	5	305	18605
66	8	528	34848
71	0	0	0
76	4	306	23104
81	5	405	32805
86	1	86	7396
	25	1740	123030

$$s^2 = \frac{\sum f \cdot X_m^2 - \frac{(\sum f \cdot X_m)^2}{n}}{n-1} = \frac{123030 - \frac{1740^2}{25}}{25-1} = \frac{1926}{24} = 80.3$$

$$s = \sqrt{80.25} = 9.0$$

27.

X_m	f	$f \cdot X_m$	$f \cdot X_m^2$
27	5	135	3645
30	9	270	8100
33	32	1056	34848
36	30	720	25920
39	12	468	18252
62	2	84	3528
	80	2733	94293

$$s^2 = \frac{\sum f \cdot X_m^2 - \frac{(\sum f \cdot X_m)^2}{n}}{n-1} = \frac{94293 - \frac{2733^2}{80}}{80-1} = \frac{926.89}{79} = 11.7$$

$$s = \sqrt{11.7} = 3.4$$

29.
C. Var $= \frac{s}{\overline{X}} = \frac{4,000}{40,000} = 0.10 = 10\%$

C. Var $= \frac{s}{\overline{X}} = \frac{2,000}{20,000} = 0.10 = 10\%$
They are equal.

31.
C. Var $= \frac{s}{\overline{X}} = \frac{6}{26} = 0.231 = 23.1\%$

C. Var $= \frac{s}{\overline{X}} = \frac{4000}{31,000} = 0.129 = 12.9\%$
The age is more variable.

33.
a. $1 - \frac{1}{5^2} = 0.96$ or 96%

b. $1 - \frac{1}{4^2} = 0.9375$ or 93.75%

35.
$\overline{X} = 5.02$ s = 0.09
At least 75% of the data values will fall withing two standard deviations of the mean; hence,
2($0.09) = $0.18 and $5.02 - $0.18 = $4.84 and $5.02 + $0.18 = $5.20. Hence at least 75% of the data values will fall between $4.84 and $5.20.

37.

$\overline{X} = 95 \quad s = 2$

At least 88.89% of the data values will fall within 3 standard deviations of the mean, hence
$95 - 3(2) = 89$ and $95 + 3(2) = 101$. Therefore at least 88.89% of the data values will fall between
89 mg and 101 mg.

39.

$\overline{X} = 12 \quad s = 3$

$20 - 12 = 8$ and $8 \div 3 = 2.67$

Hence, $1 - \frac{1}{k^2} = 1 - \frac{1}{2.67^2} = 1 - 0.14 = 0.86 = 86\%$

At least 86% of the data values will fall between 4 and 20.

41.

$26.8 + 1(4.2) = 31$

By the Empirical Rule, 68% of consumption is within 1 standard deviation of the mean. Then $\frac{1}{2}$ of 32%, or 16%, of consumption would be more than 31 pounds of citrus fruit per year.

43.

$n = 30 \quad \overline{X} = 214.97 \quad s = 20.76$ At least 75% of the data values will fall between $\overline{X} \pm 2s$.

$\overline{X} - 2(20.76) = 214.97 - 41.52 = 173.45$ and $\overline{X} + 2(20.76) = 214.97 + 41.52 = 256.49$

In this case all 30 values fall within this range; hence Chebyshev's Theorem is correct for this example.

45.

For $k = 1.5$, $1 - \frac{1}{1.5^2} = 1 - 0.44 = 0.56$ or 56%

For $k = 2$, $1 - \frac{1}{2^2} = 1 - 0.25 = 0.75$ or 75%

For $k = 2.5$, $1 - \frac{1}{2.5^2} = 1 - 0.16 = 0.84$ or 84%

For $k = 3$, $1 - \frac{1}{3^2} = 1 - 0.1111 = .8889$ or 88.89%

For $k = 3.5$, $1 - \frac{1}{3.5^2} = 1 - 0.08 = 0.92$ or 92%

47.

$\overline{X} = 13.3$

Mean Dev $= \frac{|5-13.3|+|9-13.3|+|10-13.3|+|11-13.3|+|11-13.3|}{10}$

$+ \frac{|12-13.3|+|15-13.3|+|18-13.3|+|20-13.3|+|22-13.3|}{10} = 4.36$

EXERCISE SET 3-4

1.

A z score tells how many standard deviations the data value is above or below the mean.

3.

A percentile is a relative measure while a percent is an absolute measure of the part to the total.

5.

$Q_1 = P_{25}, \quad Q_2 = P_{50}, \quad Q_3 = P_{75}$

7.

$D_1 = P_{10}, \quad D_2 = P_{20}, \quad D_3 = P_{30}$, etc

9.

a. $z = \frac{X - \overline{X}}{s} = \frac{115 - 100}{10} = 1.5$

9. continued

b. $z = \frac{124-100}{10} = 2.4$

c. $z = \frac{93-100}{10} = -0.7$

d. $z = \frac{100-100}{10} = 0$

e. $z = \frac{85-100}{10} = -1.5$

11.
a. $z = \frac{X-\overline{X}}{s} = \frac{87-84}{4} = 0.75$

b. $z = \frac{79-84}{4} = -1.25$

c. $z = \frac{93-84}{4} = 2.25$

d. $z = \frac{76-84}{4} = -2$

e. $z = \frac{82-84}{4} = -0.5$

13.
a. $z = \frac{43-40}{3} = 1$

b. $z = \frac{75-72}{5} = 0.6$

The grade in part a is higher.

15.
a. $z = \frac{3.2-4.6}{1.5} = -0.93$ b. $z = \frac{630-800}{200} = -0.85$ c. $z = \frac{43-50}{5} = -1.4$

The score in part b is the highest.

17.
a. 21st b. 58th c. 77th d. 29th

18.
a. 7   b. 25   c. 64   d. 76   e. 93

19.
a. a. 235   b. 255   c. 261   d. 275   e. 283

20.
a. 376   b. 389   c. 432   d. 473   e. 498

21.
a. 17th b. 39th c. 53rd d. 79th e. 91st

23.
$c = \frac{6(30)}{100} = 1.8$ or 2   82

25.
$c = \frac{n \cdot p}{100} = \frac{7(60)}{100} = 4.2$ or 5   Hence, 47 is the closest value to the 60th percentile.

27.

$c = \frac{6(33)}{100} = 1.98$ or 2   5, 12, 15, 16, 20, 21

                            $\uparrow P_{33}$

29.

a. 5, 12, 16, 25, 32, 38 $Q_1 = 12$, $Q_2 = 20.5$, $Q_3 = 32$

Midquartile $= \frac{12+32}{2} = 22$ Interquartile range: $32 - 12 = 20$

b. 53, 62, 78, 94, 96, 99, 103 $Q_1 = 62$, $Q_2 = 94$, $Q_3 = 99$

Midquartile $= \frac{62+99}{2} = 80.5$ Interquartile range: $99 - 62 = 37$

EXERCISE SET 3-5

1. Data arranged in order: 6, 8, 12, 19, 27, 32, 54

Minimum: 6
Q_1: 8
Median: 19
Q_3: 32
Maximum: 54
Interquartile Range: $32 - 8 = 24$

3. Data arranged in order: 188, 192, 316, 362, 437, 589

Minimum: 188
Q_1: 192
Median: $\frac{316+362}{2} = 339$
Q_3: 437
Maximum: 589
Interquartile Range: $437 - 192 = 245$

5. Data arranged in order: 14.6, 15.5, 16.3, 18.2, 19.8

Minimum: 14.6
Q_1: $\frac{14.6+15.5}{2} = 15.05$
Median: 16.3
Q_3: $\frac{18.2+19.8}{2} = 19.0$
Maximum: 19.8
Interquartile Range: 3.95

7. Minimum: 3
Q_1: 5
Median: 8
Q_3: 9
Maximum: 11
Interquartile Range: $9 - 5 = 4$

9. Minimum: 55
Q_1: 65
Median: 70
Q_3: 90
Maximum: 95

9. continued
Interquartile Range: $90 - 65 = 25$

11.
$MD = \frac{3.9+4.7}{2} = 4.3$
$Q_1 = 2.0 \quad Q_3 = 7.6$

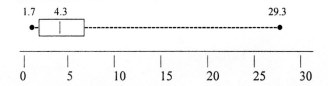

The distribution is positively skewed.

13. Data arranged in order: 13, 25, 25, 26, 28, 34, 35, 37, 42
Minimum: 13   Maximum: 42
$MD = 28$
$Q_1 = \frac{25+25}{2} = 25 \quad Q_3 = \frac{35+37}{2} = 36$

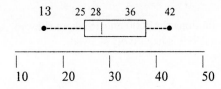

15. Data arranged in order: 3.2, 3.9, 4.4, 8.0, 9.8, 11.7, 13.9, 15.9, 17.6, 21.7, 24.8, 34.1
Minimum: 3.2   Maximum: 34.1

$MD: \frac{11.7+13.9}{2} = 12.8$

$Q_1: \frac{4.4+8.0}{2} = 6.2 \qquad Q_3: \frac{17.6+21.7}{2} = 19.65$

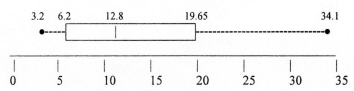

The distribution is positively skewed.

17.
(a)
For April: $\overline{X} = 149.3$
For May: $\overline{X} = 264.3$
For June: $\overline{X} = 224.0$
For July: $\overline{X} = 123.3$

The month with the highest mean number of tornadoes is May.

(b)
For 2001: $\overline{X} = 186.0$
For 2000: $\overline{X} = 165.0$
For 1999: $\overline{X} = 219.75$

17. continued
The year with the highest mean number of tornadoes is 1999.

(c) The 5-number summaries for each year are:

For 2001: 120, 127.5, 188, 244.5, 248
For 2000: 135, 135.5, 142, 194.5, 241
For 1999: 102, 139.5, 233, 300, 311

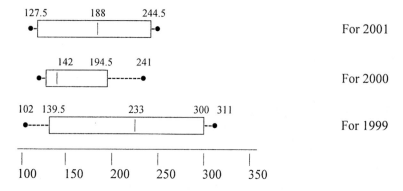

For 2001

For 2000

For 1999

The distribution for 2001 is approximately symmetric while the distributions for 2000 and 1999 are skewed.
The distribution for 2000 is positively skewed and the distribution for 1999 is negatively skewed.
2000 is the least variable and has the smallest median.

REVIEW EXERCISES - CHAPTER THREE

1.

a. $\overline{X} = \frac{\sum X}{n} = \frac{2343+1240+1088+600+497+1925+1480+458}{8} = 1203.9$

b. 458 497 600 1088 1240 1480 1925 2343
$$\uparrow$$
$$MD = \frac{1088+1240}{2} = 1164$$

c. no mode

d. $MR = \frac{458+2343}{2} = 1400.5$

e. Range $= 2343 - 458 = 1885$

f. $s^2 = \frac{\sum X^2 - \frac{(\sum X)^2}{n}}{n-1} = \frac{14,923,791 - \frac{9631^2}{8}}{8-1} = 475,610.1$

g. $s = \sqrt{475,610.1} = 689.6$

3.

Class	X_m	f	$f \cdot X_m$	$f \cdot X_m^2$	cf
1 - 3	2	1	2	4	1
4 - 6	5	4	20	100	5
7 - 9	8	5	40	320	10
10 - 12	11	1	11	121	11
13 - 15	14	1	14	196	12
		12	87	741	

3. continued

a. $\overline{X} = \dfrac{\sum f \cdot X_m}{n} = \dfrac{87}{12} = 7.3$

b. Modal Class $= 7 - 9$ or $6.5 - 9.5$

c. $s^2 = \dfrac{741 - \frac{87^2}{12}}{11} = \dfrac{110.25}{11} = 10.0$

f. $s = \sqrt{10.0} = 3.2$

5.

Class Boundaries	X_m	f	$f \cdot X_m$	$f \cdot X_m^2$	cf
12.5 - 27.5	20	6	120	2400	6
27.5 - 42.5	35	3	105	3675	9
42.5 - 57.5	50	5	250	12,500	14
57.5 - 72.5	65	8	520	33,800	22
72.5 - 87.5	80	6	480	38,400	28
87.5 - 102.5	95	2	190	18,050	30
		30	1665	108,825	

a. $\overline{X} = \dfrac{\sum f \cdot X_m}{n} = \dfrac{1665}{30} = 55.5$

b. Modal class $= 57.5 - 72.5$

c. $s^2 = \dfrac{\sum f \cdot X_m^2 - \frac{(\sum f \cdot X_m)^2}{n}}{n - 1} = \dfrac{108825 - \frac{1665^2}{30}}{30 - 1} = \dfrac{16417.5}{29} = 566.1$

d. $s = \sqrt{566.1} = 23.8$

7.

$\overline{X} = \dfrac{\sum w \cdot X}{\sum w} = \dfrac{12 \cdot 0 + 8 \cdot 1 + 5 \cdot 2 + 5 \cdot 3}{12 + 8 + 5 + 5} = \dfrac{33}{30} = 1.1$

9.

$\overline{X} = \dfrac{\sum w \cdot X}{\sum w} = \dfrac{8 \cdot 3 + 1 \cdot 6 + 1 \cdot 30}{8 + 1 + 1} = \dfrac{60}{10} = 6$

11.

Magazines: C. Var $= \dfrac{s}{\overline{X}} = \dfrac{12}{56} = 0.214$

Year: C. Var $= \dfrac{s}{\overline{X}} = \dfrac{2.5}{6} = 0.417$

The number of years is more variable.

13.

a.

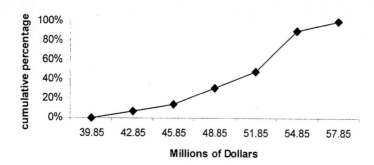

b. $P_{35} = 49$; $P_{65} = 52$; $P_{85} = 53$ (answers are approximate)

c. $44 \Rightarrow 15^{th}$ percentile; $48 \Rightarrow 33^{th}$ percentile; $54 \Rightarrow 91^{nd}$ percentile

15.

$\overline{X} = 0.32 \quad s = 0.03 \quad k = 2$

$0.32 - 2(0.03) = 0.26$ and $0.32 + 2(0.03) = 0.38$

At least 75% of the values will fall between \$0.26 and \$0.38.

17.

$\overline{X} = 54 \quad s = 4 \quad 60 - 54 = 6 \quad k = \frac{6}{4} = 1.5 \quad 1 - \frac{1}{1.5^2} = 1 - 0.44 = 0.56$ or 56%

19.

$\overline{X} = 32 \quad s = 4 \quad 44 - 32 = 12 \quad k = \frac{12}{4} = 3 \quad 1 - \frac{1}{3^2} = 0.8889 = 88.89\%$

21.

Before Christmas:

$MD = 30 \quad Q_1 = 21 \quad Q_3 = 33.5$

After Christmas:

$MD = 18 \quad Q_1 = 14.5 \quad Q_3 = 23$

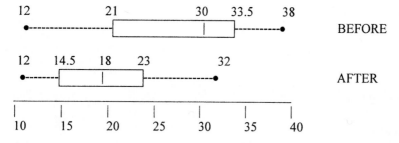

The employees worked more hours before Christmas than after Christmas. Also, the range and variability of the distribution of hours worked before Christmas is greater than that of hours worked after Christmas.

CHAPTER 3 QUIZ

1. True
2. True
3. False
4. False
5. False

6. False
7. False
8. False
9. False
10. c.
11. c.
12. a. and b.
13. b.
14. d.
15. b.
16. statistic
17. parameters, statistics
18. standard deviation
19. σ
20. midrange
21. positively
22. outlier
23. a. 84.1   b. 85   c. none   d. 84   e. 12   f. 17.1   g. 4.1
24. a. 6.4   b. 5.5 - 8.5   c. 11.6   d. 3.4
25. a. 51.4   b. 35.5 - 50.5   c. 451.5   d. 21.2
26. a. 8.2   b. 6.5 - 9.5   c. 21.6   d. 4.6
27. 1.6
28. 4.46
29. 0.33; 0.162; newspapers
30. 0.3125; 0.229; brands
31. -0.75; -1.67; science
32. a. 0.5   b. 1.6   c. 15, c is higher
33. a. 6; 19; 31; 44; 56; 69; 81; 94   b. 27

c.

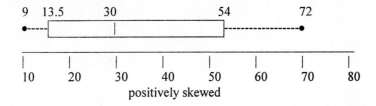

positively skewed

34.
a.

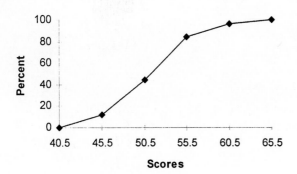

b. 47; 53; 65
c. 60th percentile; 6th percentile; 98th percentile

35.
For Pre-buy:
MD = 1.625 $Q_1 = 1.54$ $Q_3 = 1.65$

For No Pre-buy:
MD = 3.95 $Q_1 = 3.85$ $Q_3 = 3.99$

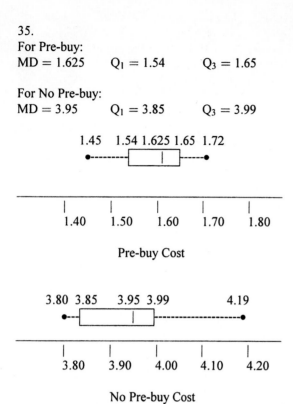

Pre-buy Cost

No Pre-buy Cost

The cost of pre-buy gas is much less than to return the car without filling it with gas. The variability of the return without filling with gas is larger than the variability of the pre-buy gas.

36.
For above 1129: 16%
For above 799: 97.5%

Note: Answers may vary due to rounding, TI-83's or computer programs.

EXERCISE 4-2

1.
A probability experiment is a chance process which leads to well-defined outcomes.

3.
An outcome is the result of a single trial of a probability experiment, whereas an event can consist of one or more outcomes.

5.
The range of values is $0 \le P(E) \le 1$.

7.
0

9.
$1 - 0.85 = 0.15$

11.
a. empirical   e. empirical
b. classical   f. empirical
c. empirical   g. subjective
d. classical

12.
a. $\frac{1}{6}$ e. 1
b. $\frac{1}{2}$ f. $\frac{5}{6}$
c. $\frac{1}{3}$ g. $\frac{1}{6}$
d. 1

13.
There are 6^2 or 36 outcomes.
a. There are 5 ways to get a sum of 6. They are (1,5), (2,4), (3,3), (4,2), and (5,1). The probability then is $\frac{5}{36}$.

b. There are six ways to get doubles. They are (1,1), (2,2), (3,3), (4,4), (5,5), and (6,6). The probability then is $\frac{6}{36} = \frac{1}{6}$.

c. There are six ways to get a sum of 7. They are (1,6), (2,5), (3,4), (4,3), (5,2), and (6,1). There are two ways to get a sum of 11. They are (5,6) and (6,5). Hence, the total number of ways to get a 7 or 11 is eight. The probability then is $\frac{8}{36} = \frac{2}{9}$.

d. To get a sum greater than nine, one must roll a 10, 11, or 12. There are six ways to

13d. continued
get a 10, 11, or 12. They are (4,6), (5,5), (6,4), (5,6), (6,5), and (6,6). The probability then is $\frac{6}{36} = \frac{1}{6}$.

e. To get a sum less than or equal to four, one must roll a 4, 3, or 2. There are six ways to do this. They are (3,1), (2,2), (1,3), (2,1), (1,2), and (1,1). The probability is $\frac{6}{36} = \frac{1}{6}$.

14.
a. $\frac{1}{13}$ f. $\frac{4}{13}$
b. $\frac{1}{4}$ g. $\frac{1}{2}$
c. $\frac{1}{52}$ h. $\frac{1}{26}$
d. $\frac{2}{13}$ i. $\frac{7}{13}$
e. $\frac{4}{13}$ j. $\frac{1}{26}$

15.
There are 24 possible outcomes.

(a) P(winning $10) = P(rolling a 1)
P(rolling a 1) $= \frac{4}{24} = \frac{1}{6}$

(b) P(winning $5 or $10) = P(rolling either a 1 or 2)
P(1 or 2) $= \frac{12}{24} = \frac{1}{2}$

(c) P(winning a coupon) = P(rolling either a 3 or 4)
P(3 or 4) $= \frac{12}{24} = \frac{1}{2}$

17.
(a) P(graduate school) $= \frac{110}{250} = \frac{11}{25}$ or 0.44

(b) P(medical school) $= \frac{10}{250} = \frac{1}{25}$ or 0.04

(c) P(not going to graduate school) $= 1 - \frac{110}{250} = \frac{140}{250}$ or 0.56

19.
(a) P(student) $= \frac{12}{18} = \frac{2}{3}$

(b) P(senior or junior) $= \frac{8}{18} = \frac{4}{9}$

(c) P(not a student) $= 1 - \frac{2}{3} = \frac{1}{3}$
or: P(not a student) = P(faculty or administrator) $= \frac{6}{18} = \frac{1}{3}$

21.

The sample space is BBB, BBG, BGB, GBB, GGB, GBG, BGG, and GGG.

a. All boys is the outcome BBB; hence P(all boys) $= \frac{1}{8}$.

b. All girls or all boys would be BBB and GGG; hence, P(all girls or all boys) $= \frac{1}{4}$.

c. Exactly two boys or two girls would be BBG, BGB, GBB, BBG, GBG, or BGG. The probability then is $\frac{6}{8} = \frac{3}{4}$.

d. At least one child of each gender means at least one boy or at least one girl. The outcomes are the same as those of part c, hence the probability is the same, $\frac{3}{4}$.

23.

The outcomes for 2, 3, or 12 are (1,1), (1,2), (2,1), and (6,6); hence P(2, 3, or 12) $= \frac{1+2+1}{36} = \frac{4}{36} = \frac{1}{9}$.

25.

a. There are 18 odd numbers; hence, P(odd) $= \frac{18}{36} = \frac{9}{19}$.

b. There are 11 numbers greater than 25 (26 through 36) hence, the probability is $\frac{11}{38}$.

c. There are 14 numbers less than 15 hence the probability is $\frac{14}{38} = \frac{7}{19}$.

27.

P(right amount or too little) $= 0.35 + 0.19$
P(right amount or too little) $= 0.54$

29.

(a)

	1	2	3	4	5	6
1	1	2	3	4	5	6
2	2	4	6	8	10	12
3	3	6	9	12	15	18
4	4	8	12	16	20	24
5	5	10	15	20	25	30
6	6	12	18	24	30	36

(b) P(multiple of 6) $= \frac{15}{36} = \frac{5}{12}$

(c) P(less than 10) $= \frac{17}{36}$

31.

a. 0.08

b. 0.01

c. $0.08 + 0.27 = 0.35$

d. $0.01 + 0.24 + 0.11 = 0.36$

33.

The statement is probably not based on empirical probability and probably not true.

35.

Actual outcomes will vary, however each number should occur approximately $\frac{1}{6}$ of the time.

37.

a. 1:5, 5:1   e. 1:12, 12:1

b. 1:1, 1:1   f. 1:3, 3:1

c. 1:3, 3:1   g. 1:1, 1:1

d. 1:1, 1:1

EXERCISE SET 4-3

1.

Two events are mutually exclusive if they cannot occur at the same time. Examples will vary.

3.

$\frac{2}{12} = \frac{1}{6}$

5.

$\frac{4}{19} + \frac{7}{19} = \frac{11}{19}$

7.

a. $\frac{5}{17} + \frac{3}{17} = \frac{8}{17}$

b. $\frac{4}{17} + \frac{2}{17} = \frac{6}{17}$

c. $\frac{3}{17} + \frac{2}{17} + \frac{4}{17} = \frac{9}{17}$

d. $\frac{5}{17} + \frac{4}{17} + \frac{3}{17} = \frac{12}{17}$

9.

P(football or basketball) $=$
$\frac{58 + 40 - 8}{200} = \frac{90}{200}$ or 0.45

P(neither) $= 1 - \frac{90}{200} = \frac{11}{20}$ or 0.55

11.

	Junior	Senior	Total
Female	6	6	12
Male	12	4	16
Total	18	10	28

a. $\frac{18}{28} + \frac{12}{28} - \frac{6}{28} = \frac{24}{28} = \frac{6}{7}$

11. continued

b. $\frac{10}{28} + \frac{12}{28} - \frac{6}{28} = \frac{16}{28} = \frac{4}{7}$

c. $\frac{18}{28} + \frac{10}{28} = \frac{28}{28} = 1$

13.

	SUV	Compact	Mid-sized	Total
Foreign	20	50	20	90
Domestic	65	100	45	210
Total	85	150	65	300

(a) P(domestic) $= \frac{210}{300} = \frac{7}{10}$

(b) P(foreign and mid-sized) $= \frac{20}{300} = \frac{1}{15} = 0.0667$

(c) P(domestic or SUV) $= \frac{210}{300} + \frac{85}{300} - \frac{65}{300}$
$= \frac{230}{300} = \frac{23}{30}$ or 0.7667

15.

	Cashier	Clerk	Deli	Total
Married	8	12	3	23
Not Married	5	15	2	22
Total	13	27	5	45

a. P(stock clerk or married) = P(clerk) + P(married) − P(married stock clerk) =
$\frac{27}{45} + \frac{23}{45} - \frac{12}{45} = \frac{38}{45}$

b. P(not married) $= \frac{22}{45}$

c. P(cashier or not married) = P(cashier) + P(not married) − P(unmarried cashier)
$= \frac{13}{45} + \frac{22}{45} - \frac{5}{45} = \frac{30}{45} = \frac{2}{3}$

17.

	Ch. 6	Ch. 8	Ch. 10	Total
Quiz	5	2	1	8
Comedy	3	2	8	13
Drama	4	4	2	10
Total	12	8	11	31

a. P(quiz show or channel 8) = P(quiz) + P(channel 8) − P(quiz show on ch. 8) =
$\frac{8}{31} + \frac{8}{31} - \frac{2}{31} = \frac{14}{31}$

b. P(drama or comedy) = P(drama) + P(comedy) $= \frac{13}{31} + \frac{10}{31} = \frac{23}{31}$

c. P(channel 10 or drama) = P(ch. 10) + P(drama) − P(drama on channel 10) =
$\frac{11}{31} + \frac{10}{31} - \frac{2}{31} = \frac{19}{31}$

19.
The total of the frequencies is 30.

a. $\frac{2}{30} = \frac{1}{15}$

b. $\frac{2+3+5}{30} = \frac{10}{30} = \frac{1}{3}$

c. $\frac{12+8+2+3}{30} = \frac{25}{30} = \frac{5}{6}$

d. $\frac{12+8+2+3}{30} = \frac{25}{30} = \frac{5}{6}$

e. $\frac{8+2}{30} = \frac{10}{30} = \frac{1}{3}$

21.
The total of the frequencies is 24.

a. $\frac{10}{24} = \frac{5}{12}$

b. $\frac{2+1}{24} = \frac{3}{24} = \frac{1}{8}$

c. $\frac{10+3+2+1}{24} = \frac{16}{24} = \frac{2}{3}$

d. $\frac{8+10+3+2}{24} = \frac{23}{24}$

23.
a. There are 4 kings, 4 queens, and 4 jacks; hence P(king or queen or jack) $= \frac{12}{52} = \frac{3}{13}$

b. There are 13 clubs, 13 hearts, and 13 spades; hence, P(club or heart or spade) = $\frac{13+13+13}{52} = \frac{39}{52} = \frac{3}{4}$

c. There are 4 kings, 4 queens, and 13 diamonds but the king and queen of diamonds were counted twice, hence; P(king or queen or diamond) = P(king) + P(queen) + P(diamond) − P(king and queen of diamonds) $= \frac{4}{52} + \frac{4}{52} + \frac{13}{52} - \frac{2}{52} = \frac{19}{52}$

d. There are 4 aces, 13 diamonds, and 13 hearts. There is one ace of diamonds and one ace of hearts; hence, P(ace or diamond or heart) = P(ace) + P(diamond) + P(heart) − P(ace of hearts and ace of diamonds) = $\frac{4}{52} + \frac{13}{52} + \frac{13}{52} - \frac{2}{52} = \frac{28}{52} = \frac{7}{13}$

e. There are 4 nines, 4 tens, 13 spades, and 13 clubs. There is one nine of spades, one ten of spades, one nine of clubs and one ten of clubs. Hence, P(9 or 10 or spade or club) = P(9) + P(10) + P(spade) + P(club) − P(9 and 10 of clubs and spades) = $\frac{4}{52} + \frac{4}{52} + \frac{13}{52} + \frac{13}{52} - \frac{4}{52} = \frac{30}{52} = \frac{15}{26}$

25.

$P(\text{red or white ball}) = \frac{7}{10}$

27.

$P(\text{mushrooms or pepperoni}) =$
$\qquad P(\text{mushrooms}) + P(\text{pepperoni}) -$
$\qquad P(\text{mushrooms and pepperoni})$
Let $X = P(\text{mushrooms and pepperoni})$
Then $0.55 = 0.32 + 0.17 - X$
$X = 0.06$

29.

$P(\text{not a two-car garage}) = 1 - 0.70 = 0.30$

EXERCISE SET 4-4

1.

a. independent   e. independent
b. dependent   f. dependent
c. dependent   g. dependent
d. dependent   h. independent

3.

$P(\text{two with elevated blood pressure}) =$
$(.68)^2 = 0.462$ or 46.2%

5.

$P(\text{5 are tippers}) = (0.73)^5 = 0.2073$

7.

(a) $P(\text{no computer}) = 1 - 0.543 = 0.457$
$P(\text{none of three has a computer}) =$
$(0.457)^3 = 0.0954$

(b) $P(\text{at least one has a computer}) =$
$1 - P(\text{none of three has a computer}) =$
$1 - 0.0954 = 0.9054$

(c) $P(\text{all three have computers}) =$
$(0.543)^3 = 0.1601$

9.

$P(\text{all are citizens}) = (0.801)^3 = 0.5139$

11.

$P(\text{all three have NFL apparel}) =$
$(0.31)^3 = 0.0298$

13.

$P(\text{no insurance}) = 0.12$
$P(\text{none are covered}) = (0.12)^4 = 0.0002$

15.

$P(\text{5 buy at least 1}) = (\frac{90}{120})^5 = \frac{243}{1024}$

17.

$\frac{5}{8} \cdot \frac{4}{7} \cdot \frac{3}{6} = \frac{5}{28}$

19.

$\frac{18}{30} \cdot \frac{17}{29} = \frac{51}{145}$

21.

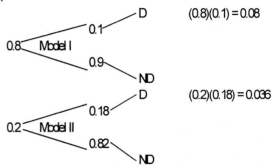

$P(\text{defective}) = 0.08 + 0.036 = 0.116$

23.

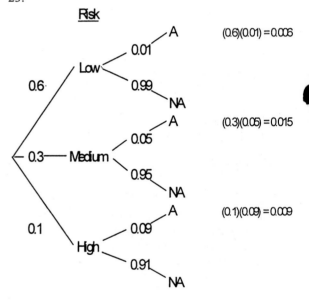

$P(\text{accident}) = .006 + .015 + .009 = 0.03$

25.

$P(\text{red ball}) = \frac{1}{3} \cdot \frac{5}{8} + \frac{1}{3} \cdot \frac{3}{4} + \frac{1}{3} \cdot \frac{4}{6} = \frac{49}{72}$

27.

$P(\text{auto will be found within one week} \mid \text{it's}$
$\text{been stolen}) = \frac{P(\text{stolen and found within 1 week})}{P(\text{stolen})}$
$= \frac{0.0009}{0.0015} = 0.6$

29.

$P(\text{swim} \mid \text{bridge}) = \frac{P(\text{play bridge and swim})}{P(\text{play bridge})}$

29. continued

$P = \frac{0.73}{0.82} = 0.89$ or 89%

31.

P(garage | deck) $= \frac{0.42}{0.60} = 0.7$ or 70%

33.

P(champagne | bridge) $= \frac{0.68}{0.83} = 0.82$ or 82%

35.

(a) P(foreign patent | corporation) =

$\frac{\text{P(corporation and foreign patent)}}{\text{P(corporation)}} =$

$\frac{\frac{63,182}{147,497}}{\frac{134,076}{147,497}} = \frac{63,182}{134,076} = 0.4712$

(b) P(individual | U. S.) $= \frac{\text{P(U. S. \& individual)}}{\text{P(U. S.)}}$

$\frac{\frac{6129}{147,497}}{\frac{77,944}{147,497}} = \frac{6129}{77,944} = 0.0786$

37.

(a) P(none have been married) $= (0.703)^5 = 0.1717$

(b) P(at least one has been married) = 1 − P(none have been married)
$= 1 - 0.1717$
$= 0.8283$

39.

P(at least one not immunized) = 1 − P(none of the six are not immunized)
$= 1 - $ P(all six are immunized)
$= 1 - (0.76)^6 = 0.8073$

41.

If P(read to) $= 0.58$, then
P(not being read to) $= 1 - 0.58 = 0.42$
P(at least one is read to) $= 1 - $ P(none are read to)
$= 1 - $ P(all five are not read to)
$= 1 - (0.42)^5 = 0.9869$

43.

P(at least one club) = 1 − P(no clubs)
$1 - \frac{39}{52} \cdot \frac{38}{51} \cdot \frac{37}{50} \cdot \frac{36}{49} = 1 - \frac{6327}{20,825}$
$= \frac{14,498}{20,825}$

45.

P(at least one defective) = 1 − P(no defective) $= 1 - (.94)^5 = 0.266$ or 26.6%

47.

P(at least one tail) = 1 − P(no tails)
$1 - (\frac{1}{2})^6 = 1 - \frac{1}{64} = \frac{63}{64}$

49.

P(at least one 6) = 1 − P(no 6's)
$1 - (\frac{5}{6})^5 = 1 - \frac{3125}{7776} = \frac{4651}{7776}$

51.

P(at least one even) = 1 − P(no evens)
$1 - (\frac{1}{2})^3 = 1 - \frac{1}{8} = \frac{7}{8}$

53.

No, because P(A ∩ B) = 0 therefore
P(A ∩ B) ≠ P(A) · P(B)

55.

P(enroll) = 0.55

P(enroll | DW) > P(enroll) which indicates that DW has a positive effect on enrollment.

P(enroll | LP) = P(enroll) which indicates that LP has no effect on enrollment.

P(enroll | MH) < P(enroll) which indicates that MH has a detrimental effect on enrollment.

Thus, all students should meet with DW.

EXERCISE SET 4-5

1.
$10^5 = 100,000$
$10 \cdot 9 \cdot 8 \cdot 7 \cdot 6 = 30,240$

3.
$7! = 7 \cdot 6 \cdot 5 \cdot 4 \cdot 3 \cdot 2 \cdot 1 = 5040$

5.
$8! = 8 \cdot 7 \cdot 6 \cdot 5 \cdot 4 \cdot 3 \cdot 2 \cdot 1 = 40,320$

7.
$5! = 5 \cdot 4 \cdot 3 \cdot 2 \cdot 1 = 120$

9.
$10 \cdot 10 \cdot 10 = 1000$
$1 \cdot 9 \cdot 8 = 72$

11.

$5 \cdot 2 = 10$

13.

(a) $8! = 8 \cdot 7 \cdot 6 \cdot 5 \cdot 4 \cdot 3 \cdot 2 \cdot 1 = 40,320$

(b) $10! = 10 \cdot 9 \cdot 8 \cdot 7 \cdot 6 \cdot 5 \cdot 4 \cdot 3 \cdot 2 \cdot 1$
$10! = 3,628,800$

(c) $0! = 1$

(d) $1! = 1$

(e) $_7P_5 = \frac{7!}{(7-5)!}$

$= \frac{7 \cdot 6 \cdot 5 \cdot 4 \cdot 3 \cdot 2 \cdot 1}{2 \cdot 1} = 2520$

(f) $_{12}P_4 = \frac{12!}{(12-4)!}$

$= \frac{12 \cdot 11 \cdot 10 \cdot 9 \cdot 8 \cdot 7 \cdot 6 \cdot 5 \cdot 4 \cdot 3 \cdot 2 \cdot 1}{8 \cdot 7 \cdot 6 \cdot 5 \cdot 4 \cdot 3 \cdot 2 \cdot 1} = 11,880$

(g) $_5P_3 = \frac{5!}{(5-3)!}$

$= \frac{5 \cdot 4 \cdot 3 \cdot 2 \cdot 1}{2 \cdot 1} = 60$

(h) $_6P_0 = \frac{6!}{(6-0)!}$

$= \frac{6 \cdot 5 \cdot 4 \cdot 3 \cdot 2 \cdot 1}{6 \cdot 5 \cdot 4 \cdot 3 \cdot 2 \cdot 1} = 1$

(i) $_5P_5 = \frac{5!}{(5-5)!}$

$= \frac{5 \cdot 4 \cdot 3 \cdot 2 \cdot 1}{0!} = 120$

(j) $_6P_2 = \frac{6!}{(6-2)!}$

$= \frac{6 \cdot 5 \cdot 4 \cdot 3 \cdot 2 \cdot 1}{4 \cdot 3 \cdot 2 \cdot 1} = 30$

15.

$_4P_4 = \frac{4!}{(4-4)!} = \frac{4 \cdot 3 \cdot 2 \cdot 1}{0!} = 24$

17.

$_6P_3 = \frac{6!}{(6-3)!} = \frac{6!}{3!} = 120$

19.

$_7P_4 = \frac{7!}{(7-4)!} = \frac{7 \cdot 6 \cdot 5 \cdot 4 \cdot 3 \cdot 2 \cdot 1}{3 \cdot 2 \cdot 1} = 840$

21.

$_{10}P_6 = \frac{10!}{(10-6)!} = \frac{10 \cdot 9 \cdot 8 \cdot 7 \cdot 6 \cdot 5 \cdot 4 \cdot 3 \cdot 2 \cdot 1}{4 \cdot 3 \cdot 2 \cdot 1} = 151,200$

23.

$_{50}P_4 = \frac{50!}{(50-4)!} = \frac{50!}{46!} = 5,527,200$

25.

$_5P_3 + {_5P_4} + {_5P_5} = \frac{5!}{2!} + \frac{5!}{1!} + \frac{5!}{0!}$

$= 60 + 120 + 120 = 300$

27.

a. $\frac{5!}{3! \, 2!} = 10$ f. $\frac{3!}{3! \, 0!} = 1$

b. $\frac{8!}{5! \, 3!} = 56$ g. $\frac{3!}{0! \, 3!} = 1$

c. $\frac{7!}{3! \, 4!} = 35$ h. $\frac{9!}{2! \, 7!} = 36$

d. $\frac{6!}{4! \, 2!} = 15$ i. $\frac{12!}{10! \, 2!} = 66$

e. $\frac{6!}{2! \, 4!} = 15$ j. $\frac{4!}{1! \, 3!} = 4$

29.

$_{10}C_3 = \frac{10!}{7! \, 3!} = \frac{10 \cdot 9 \cdot 8 \cdot 7!}{7! \cdot 3 \cdot 2 \cdot 1} = 120$

31.

$_{10}C_4 = \frac{10!}{6! \, 4!} = 210$

33.

$_{20}C_5 = \frac{20!}{15! \, 5!} = 15,504$

35.

$_8C_3 \cdot {_{11}C_4} = 56 \cdot 330 = 18,480$

37.

$_{12}C_4 = 495$
$_7C_2 \cdot {_5C_2} = 21 \cdot 10 = 210$
$_7C_2 \cdot {_5C_2} + {_7C_3} \cdot {_5C_1} + {_7C_4} =$
$21 \cdot 10 + 35 \cdot 5 + 35 =$
$210 + 175 + 35 = 420$

39.

$_6C_3 \cdot {_5C_2} = \frac{6!}{3! \, 3!} \cdot \frac{5!}{3! \, 2!}$
$= \frac{6 \cdot 5 \cdot 4 \cdot 3!}{3! \cdot 3 \cdot 2 \cdot 1} \cdot \frac{5 \cdot 4 \cdot 3!}{3! \cdot 2 \cdot 1} = 200$

41.

$_{10}C_2 \cdot {_{12}C_2} = \frac{10!}{8! \, 2!} \cdot \frac{12!}{10! \, 2!}$
$= 45 \cdot 66 = 2,970$

43.

$_{17}C_2 = \frac{17!}{15! \, 2!} = 136$

45.

$_{11}C_7 = \frac{11!}{4! \, 7!} = \frac{11 \cdot 10 \cdot 9 \cdot 8 \cdot 7!}{7! \cdot 4 \cdot 3 \cdot 2 \cdot 1} = 330$

47.

$_{20}C_8 = \frac{20!}{12! \, 8!} = \frac{20 \cdot 19 \cdot 18 \cdot 17 \cdot 16 \cdot 15 \cdot 14 \cdot 13 \cdot 12!}{12! \cdot 8 \cdot 7 \cdot 6 \cdot 5 \cdot 4 \cdot 3 \cdot 2 \cdot 1}$
$= 125,970$

49.

Selecting 1 coin there are 4 ways. Selecting 2 coins there are 6 ways. Selecting 3 coins there are 4 ways. Selecting 4 coins there is 1 way. Hence the total is $4 + 6 + 4 + 1 = 15$ ways. (List all possibilities.)

51.
a. $2 \cdot 4 \cdot 3 \cdot 2 \cdot 1 = 48$
b. $4 \cdot 6 + 3 \cdot 6 + 2 \cdot 6 + 1 \cdot 6 = 60$
c. $5! - 48 = 72$

EXERCISE SET 4-6

1.
$P(\text{2 face cards}) = \frac{12}{52} \cdot \frac{11}{51} = \frac{11}{221}$

3.
a. There are $_4C_3$ ways of selecting 3 women and $_7C_3$ total ways to select 3 people; hence, $P(\text{all women}) = \frac{_4C_3}{_7C_3} = \frac{4}{35}$.

b. There are $_3C_3$ ways of selecting 3 men; hence, $P(\text{all men}) = \frac{_3C_3}{_7C_3} = \frac{1}{35}$.

c. There are $_3C_2$ ways of selecting 2 men and $_4C_1$ ways of selecting one woman; hence, $P(\text{2 men and 1 woman}) = \frac{_3C_2 \cdot _4C_1}{_7C_3}$ $= \frac{12}{35}$.

d. There are $_3C_1$ ways to select one man and $_4C_2$ ways of selecting two women; hence, $P(\text{1 man and 2 women}) = \frac{_3C_1 \cdot _4C_2}{_7C_3}$ $= \frac{18}{35}$.

5.
(a) There are $_9C_4$ ways to select four from Pennsylvania; hence $P(\text{all four are from Pennsylvania}) = \frac{_9C_4}{_{56}C_4} = \frac{126}{367,290} = 0.0003$

(b) There are $_9C_2$ ways to select two from Pennsylvania and $_7C_2$ ways to select two from Virginia; hence $P(\text{two from Pennsylvania and two from Virginia}) = \frac{_9C_2 \cdot _7C_2}{_{56}C_4} = \frac{756}{367,290} = 0.0021$

7.
$\frac{2}{50} \cdot \frac{1}{49} = \frac{1}{1225}$

9.
a. $\frac{_8C_4}{_{14}C_4} = \frac{70}{1001} = \frac{10}{143}$

9. continued

b. $\frac{_6C_2 \cdot _8C_2}{_{14}C_4} = \frac{420}{1001} = \frac{60}{143}$

c. $\frac{_6C_4}{_{14}C_4} = \frac{15}{1001}$

d. $\frac{_6C_3 \cdot _8C_1}{_{14}C_4} = \frac{160}{1001}$

e. $\frac{_6C_1 \cdot _8C_3}{_{14}C_4} = \frac{336}{1001} = \frac{48}{143}$

11.
(a) $\frac{_{11}C_2}{_{19}C_2} = \frac{55}{171} = 0.3216$

(b) $\frac{_8C_2}{_{19}C_2} = \frac{28}{171} = 0.1637$

(c) $\frac{_{11}C_1 \cdot _8C_1}{_{19}C_2} = \frac{88}{171} = 0.5146$

(d) It probably got lost in the wash!

13.
There are $6^3 = 216$ ways of tossing three dice, and there are 15 ways of getting a sum of 7; i.e., (1, 1, 5), (1, 5, 1), (5, 1, 1), (1, 2, 4), etc. Hence the probability of rolling a sum of 7 is $\frac{15}{216} = \frac{5}{72}$.

15.
There are $5! = 120$ ways to arrange 5 washers in a row and 2 ways to have them in correct order, small to large or large to small; hence, the probability is $\frac{2}{120} = \frac{1}{60}$.

REVIEW EXERCISES - CHAPTER FOUR

1.
a. $\frac{1}{6}$　　b. $\frac{1}{6}$　　c. $\frac{4}{6} = \frac{2}{3}$

3.
$\frac{16}{45}$

5.
$\frac{850}{1500} = \frac{17}{30}$

7.
a. $\frac{3}{30} = \frac{1}{10}$　　c. $\frac{16+7+3}{30} = \frac{26}{30} = \frac{13}{15}$

b. $\frac{7+4}{30} = \frac{11}{30}$　　d. $1 - \frac{4}{30} = \frac{26}{30} = \frac{13}{15}$

9.
$0.80 + 0.30 - 0.12 = 0.98$

11.
$(0.78)^5 = 0.289$ or 28.9%

13.

a. $\frac{26}{52} \cdot \frac{25}{51} \cdot \frac{24}{50} = \frac{2}{17}$

b. $\frac{13}{52} \cdot \frac{12}{51} \cdot \frac{11}{50} = \frac{33}{2550} = \frac{11}{850}$

c. $\frac{4}{52} \cdot \frac{3}{51} \cdot \frac{2}{50} = \frac{1}{5525}$

15.
P(C or PP) = P(C) + P(PP) = $\frac{2+3}{13} = \frac{5}{13}$

17.

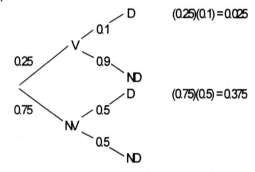

P(disease) = $0.025 + 0.375 = 0.4$

19.
P(NC | C) = $\frac{P(NC \text{ and } C)}{P(C)} = \frac{0.37}{0.73} = 0.51$

21.
$\frac{0.43}{0.75} = 0.573$ or 57.3%

23.

	<4 yrs HS	HS	College	Total
Smoker	6	14	19	39
Non-Smoker	18	7	25	50
Total	24	21	44	89

a. There are 44 college graduates and 19 of them smoke; hence, the probability is $\frac{19}{44}$.

b. There are 24 people who did not graduate from high school, 6 of whom do not smoke; hence, the probability is
$\frac{6}{24} = \frac{1}{4}$.

25.
P(at least one tail) = $1 -$ P(all heads)
$1 - \left(\frac{1}{2}\right)^5 = 1 - \frac{1}{32} = \frac{31}{32}$

27.
If repetitions are allowed:
$26 \cdot 26 \cdot 26 \cdot 10 \cdot 10 \cdot 10 = 175,760,000$

If repetitions are not allowed:
$_{26}P_3 \cdot {_{10}}P_4 = \frac{26 \cdot 25 \cdot 24 \cdot 23!}{23!} \cdot \frac{10 \cdot 9 \cdot 8 \cdot 7 \cdot 6!}{6!}$
$= 78,624,000$

If repetitions are allowed in the letters but not in the digits:
$26 \cdot 26 \cdot 26 \cdot {_{10}}P_4 = 88,583,040$

29.
$_5C_3 \cdot {_7}C_4 = \frac{5!}{2!\,3!} \cdot \frac{7!}{3!\,4!} = 10 \cdot 35 = 350$

31.
$_{10}C_2 = \frac{10!}{8!\,2!} = 45$

33.
$26 \cdot 10 \cdot 10 \cdot 10 = 26,000$

35.
$_{12}C_4 = \frac{12!}{8!\,4!} = \frac{12 \cdot 11 \cdot 10 \cdot 9 \cdot 8!}{4 \cdot 3 \cdot 2 \cdot 1 \cdot 8!} = 495$

37.
$_{20}C_5 = \frac{20!}{15!\,5!} = \frac{20 \cdot 19 \cdot 18 \cdot 17 \cdot 16 \cdot 15!}{15!\,5 \cdot 4 \cdot 3 \cdot 2 \cdot 1} = 15,504$

39.
Total number of outcomes:
$26 \cdot 26 \cdot 26 \cdot 10 \cdot 10 \cdot 10 \cdot 10 = 175,760,000$

Total number of ways for USA followed by a number divisible by 5:
$1 \cdot 1 \cdot 1 \cdot 10 \cdot 10 \cdot 10 \cdot 2 = 2000$

Hence P = $\frac{2000}{175,760,000} = 0.000011$

41.
$\frac{_3C_1 \cdot {_4}C_1 \cdot {_2}C_1}{_9C_3} = \frac{2}{7}$

CHAPTER FOUR QUIZ

1. False, subjective probability can be used when other types of probabilities cannot be found.
2. False, empirical probability uses frequency distributions.
3. True
4. False, P(A or B) = P(A) + P(B) − P(A and B)
5. False, the probabilities can be different.
6. False, complementary events cannot occur at the same time.

7. True

8. False, order does not matter in combinations.

9. b.

10. b. and d.

11. d.

12. b.

13. c.

14. b.

15. d.

16. b.

17. b.

18. sample space

19. zero and one

20. zero

21. one

22. mutually exclusive

23. a. $\frac{4}{52} = \frac{1}{13}$ c. $\frac{16}{52} = \frac{4}{13}$

    b. $\frac{4}{52} = \frac{1}{13}$

24. a. $\frac{13}{52} = \frac{1}{4}$ d. $\frac{4}{52} = \frac{1}{13}$

    b. $\frac{4+13-1}{52} = \frac{4}{13}$ e. $\frac{26}{52} = \frac{1}{2}$

    c. $\frac{1}{52}$

25. a. $\frac{12}{31}$ c. $\frac{27}{31}$

    b. $\frac{12}{31}$ d. $\frac{24}{31}$

26. a. $\frac{11}{36}$ d. $\frac{1}{3}$

    b. $\frac{5}{18}$ e. 0

    c. $\frac{11}{36}$ f. $\frac{11}{12}$

27. $0.75 + 0.25 - 0.16 = 0.84$

28. $(0.3)^5 = 0.002$

29. a. $\frac{26}{52} \cdot \frac{25}{51} \cdot \frac{24}{50} \cdot \frac{23}{49} \cdot \frac{22}{48} = \frac{253}{9996}$

    b. $\frac{13}{52} \cdot \frac{12}{51} \cdot \frac{11}{50} \cdot \frac{10}{49} \cdot \frac{9}{48} = \frac{33}{66,640}$

    c. 0

30. $\frac{0.35}{0.65} = 0.54$

31. $\frac{0.16}{0.3} = 0.53$

32. $\frac{0.57}{0.7} = 0.81$

33. $\frac{0.028}{0.5} = 0.056$

34. a. $\frac{1}{2}$ b. $\frac{3}{7}$

35. $1 - (0.45)^6 = 0.99$

36. $1 - (\frac{5}{6})^4 = 0.518$

37. $1 - (0.15)^6 = 0.9999886$

38. 2,646

39. 40,320

40. 1,365

41. 1,188,137,600; 710,424,000

42. 720

43. 33,554,432

44. 35

45. $\frac{1}{4}$

46. $\frac{3}{14}$

47. $\frac{12}{55}$

Note: Answers may vary due to rounding, TI-83's or computer programs.

EXERCISE SET 5-2

1.
A random variable is a variable whose values are determined by chance. Examples will vary.

3.
The number of commercials a radio station plays during each hour.
The number of times a student uses his or her calculator during a mathematics exam.
The number of leaves on a specific type of tree.

5.
A probability distribution is a distribution which consists of the values a random variable can assume along with the corresponding probabilities of these values.

7.
Yes

9.
Yes

11.
No, probability values cannot be greater than 1.

13.
Discrete

15.
Continuous

17.
Discrete

19.

X	0	1	2	3
P(X)	$\frac{6}{15}$	$\frac{5}{15}$	$\frac{3}{15}$	$\frac{1}{15}$

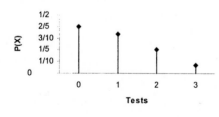

21.

X	0	1	2	3	4	5
P(X)	0.75	0.17	0.04	0.025	0.01	0.005

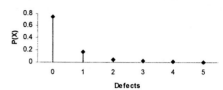

23.

X	1	2	3	4	5	6
P(X)	$\frac{1}{2}$	$\frac{1}{6}$	$\frac{1}{12}$	$\frac{1}{12}$	$\frac{1}{12}$	$\frac{1}{12}$

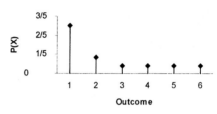

25.

X	1	2	3	4	5
P(X)	0.1	0.25	0.25	0.2	0.2

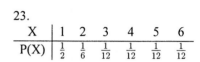

27.

X	$1	$5	$10	$20
P(X)	$\frac{3}{7}$	$\frac{2}{7}$	$\frac{1}{7}$	$\frac{1}{7}$

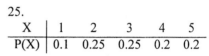

29.

X	1	2	3	4
P(X)	$\frac{1}{4}$	$\frac{1}{4}$	$\frac{3}{8}$	$\frac{1}{8}$

31.

X	1	2	3
P(X)	$\frac{1}{6}$	$\frac{1}{3}$	$\frac{1}{2}$

Yes.

33.

X	3	4	7
P(X)	$\frac{3}{6}$	$\frac{4}{6}$	$\frac{7}{6}$

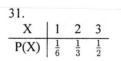
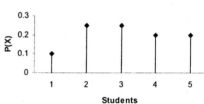

33. continued

No, the sum of the probabilities is greater than one and $P(7) = \frac{7}{6}$ which is also greater than one.

35.

X	1	2	4
P(X)	$\frac{1}{7}$	$\frac{2}{7}$	$\frac{4}{7}$

Yes.

EXERCISE SET 5-3

1.

X	0	1	2	3
P(X)	0.92	0.03	0.03	0.02

$\mu = \sum X \cdot P(X) = 0(0.92) + 1(0.03) +$

$2(0.03) + 3(0.02) = 0.15$ or 0.2

$\sigma^2 = \sum X^2 \cdot P(X) - \mu^2 = 0^2(0.92) +$

$1^2(0.03) + 2^2(0.03) + 3^2(0.02) = 0.3075$

$\sigma = \sqrt{0.3075} = 0.55$ or 0.6

The company would need $0.2(10) = 2$ extra transistors on hand each day.

X	P(X)	$X \cdot P(X)$	$X^2 \cdot P(X)$
0	0.92	0	0
1	0.03	0.03	0.03
2	0.03	0.06	0.12
3	0.02	0.06	0.18
		$\mu = 0.15$	0.33

3.

$\mu = \sum X \cdot P(X) = 0(0.18) + 1(0.44) +$

$2(0.27) + 3(0.08) + 4(0.03) = 1.34$ or 1.3

$\sigma^2 = \sum X^2 \cdot P(X) - \mu^2 = [0^2(0.18) +$

$1^2(0.44) + 2^2(0.27) + 3^2(0.08) + 4^2(0.03)]$

$- 1.34^2 = 0.92$ or 0.9

$\sigma = \sqrt{0.92} = 0.96$ or 1

No, on average each person has about one credit card.

X	P(X)	$X \cdot P(X)$	$X^2 \cdot P(X)$
0	0.18	0	0
1	0.44	0.44	0.44
2	0.27	0.54	1.08
3	0.08	0.24	0.72
4	0.03	0.12	0.48
		$\mu = 1.34$	2.72

5.

$\mu = \sum X \cdot P(X) = 0(0.06) + 1(0.42) +$

$2(0.22) + 3(0.12) + 4(0.15) + 5(0.03)$

$= 1.97$ or 2.0

$\sigma^2 = \sum X^2 \cdot P(X) - \mu^2 = [0^2(0.06) +$

$1^2(0.42) + 2^2(0.22) + 3^2(0.12) + 4^2(0.15)$

$+ 5^2(0.03)] - 1.97^2 = 1.65$ or 1.6

$\sigma = \sqrt{1.65} = 1.28$ or 1.3

X	P(X)	$X \cdot P(X)$	$X^2 \cdot P(X)$
0	0.06	0.00	0.00
1	0.42	0.42	0.42
2	0.22	0.44	0.88
3	0.12	0.36	1.08
4	0.15	0.60	2.40
5	0.03	0.15	0.75
		$\mu = 1.97$	5.53

She would average $200 per week.

7.

$\mu = \sum X \cdot P(X) = 5(0.2) + 6(0.25) +$

$7(0.38) + 8(0.10) + 9(0.07) = 6.59$ or 6.6

$\sigma^2 = \sum X^2 \cdot P(X) - \mu^2 = [5^2(0.2) +$

$6^2(0.25) + 7^2(0.38) + 8^2(0.10) +$

$9^2(0.07) - 6.59^2 = 1.2619$ or 1.3

$\sigma = \sqrt{1.2619} = {}^{\backprime}1.123$ or 1.1

X	P(X)	$X \cdot P(X)$	$X^2 \cdot P(X)$
5	0.20	1.00	5.00
6	0.25	1.50	9.00
7	0.38	2.66	18.62
8	0.10	0.80	6.40
9	0.07	0.63	5.67
		$\mu = 6.59$	44.69

9.

$\mu = \sum X \cdot P(X) = 12(0.15) + 13(0.20) +$

$14(0.38) + 15(0.18) + 16(0.09) = 13.86$

$\sigma^2 = \sum X^2 \cdot P(X) - \mu^2 = [12^2(0.15) +$

$13^2(0.20) + 14^2(0.38) + 15^2(0.18) +$

$16^2(0.09)] - 13.86^2 = 1.3204$ or 1.3

$\sigma = \sqrt{1.3204} = 1.1491$ or 1.1

9. continued

X	P(X)	X · P(X)	X² · P(X)
12	0.15	1.80	21.60
13	0.20	2.60	33.80
14	0.38	5.32	74.48
15	0.18	2.70	40.50
16	0.09	1.44	23.04
		$\mu = 13.86$	193.42

11.
$E(X) = \sum X \cdot P(X) = \$4995(\frac{1}{2500}) - \$5(\frac{2499}{2500}) = -\$3$

Alternate Solution:
$\$5000(\frac{1}{2500}) - \$5 = -\$3$

Yes, they will make $7500.

13.
$E(X) = \sum X \cdot P(X) = \$5.00(\frac{1}{6}) = \$0.83$
He should pay about $0.83.

15.
$E(X) = \sum X \cdot P(X) = \$1000(\frac{1}{1000}) + \$500(\frac{1}{1000}) + \$100(\frac{5}{1000}) - \$3.00$
$= -\$1.00$

Alternate Solution:
$E(X) = 997(\frac{1}{1000}) + 497(\frac{1}{1000}) + 97(\frac{5}{1000}) - 3(\frac{993}{1000}) = -\$1.00$

17.
$E(X) = \sum X \cdot P(X) = \$500(\frac{1}{1000}) - \$1.00$
$= -\$0.50$

If 123 is boxed:
$E(X) = \$499(\frac{1}{1000}) - 1(\frac{999}{1000}) = -\$0.50$

There are 6 possibilities when a number with all different digits is boxed, $(3 \cdot 2 \cdot 1 = 6)$. Hence,
$\$80.00 \cdot \frac{6}{1000} - \$1.00 = \$0.48 - \$1.00$
$= -\$0.52$

Alternate Solution:
$E(X) = 79(\frac{6}{1000}) - 1(\frac{994}{1000}) = -\$0.52$

19.
The probabilities of each are:
Red: $\frac{18}{38}$ Black: $\frac{18}{38}$

1 − 18: $\frac{18}{38}$ 19 − 36: $\frac{18}{38}$

19. continued
0: $\frac{1}{38}$ 00: $\frac{1}{38}$

Any single number: $\frac{1}{38}$

0 or 00: $\frac{2}{38}$

$E(X) = \sum X \cdot P(X)$

a. $\$1.00(\frac{18}{38}) - \$1.00(\frac{20}{38}) = -\$0.05$

b. $\$1.00(\frac{18}{38}) - \$1.00(\frac{20}{38}) = -\$0.05$

c. $\$35(\frac{1}{38}) - \$1.00(\frac{37}{38}) = -\$0.05$

d. $\$35(\frac{1}{38}) - \$1.00(\frac{37}{38}) = -\$0.05$

e. $\$17(\frac{2}{38}) - \$1.00(\frac{36}{38}) = -\$0.05$

21.
The expected value for a single die is 3.5, and since 3 die are rolled, the expected value is $3(3.5) = 10.5$

23.
Answers will vary.

25.
Answers will vary.

EXERCISE SET 5-4

1.
a. Yes
b. Yes
c. Yes
d. No, there are more than two outcomes.
e. No
f. Yes
g. Yes
h. Yes
i. No, there are more than two outcomes.
j. Yes

2.
a. 0.420
b. 0.346
c. 0.590
d. 0.251
e. 0.000
f. 0.250
g. 0.418
h. 0.176
i. 0.246

3.

a. $P(X) = \frac{n!}{(n-X)!\,X!} \cdot p^X \cdot q^{n-X}$

$P(X) = \frac{6!}{3! \cdot 3!} \cdot (0.03)^3 (0.97)^3 = 0.0005$

b. $P(X) = \frac{4!}{2! \cdot 2!} \cdot (0.18)^2 \cdot (0.82)^2 = 0.131$

c. $P(X) = \frac{5!}{2! \cdot 3!} = (0.63)^3 \cdot (0.37)^2 = 0.342$

d. $P(X) = \frac{9!}{9! \cdot 0!} \cdot (0.42)^0 \cdot (0.58)^9 = 0.007$

e. $P(X) = \frac{10!}{5! \cdot 5!} \cdot (0.37)^5 \cdot (0.63)^5 = 0.173$

5.

n = 10, p = 0.5, X = 6, 7, 8, 9, 10
$P(X) = 0.205 + 0.117 + 0.044 + 0.010 + 0.001 = 0.377$
No, because your score would be about 40%.

7.

n = 9, p = 0.30, X = 3
$P(X) = 0.267$

9.

n = 7, p = 0.75, X = 0, 1, 2, 3

$P(X) = \frac{7!}{7!\,0!}(0.75)^0 (0.25)^7 +$

$\frac{7!}{6!\,1!}(0.75)^1 (0.25)^6 + \frac{7!}{5!\,2!}(0.75)^2 (0.25)^5 +$

$\frac{7!}{4!\,3!}(0.75)^3 (0.25)^4 = 0.071$

11.

n = 5, p = 0.40
a. X = 2, P(X) = 0.346

b. X = 0, 1, 2, or 3 people
$P(X) = 0.078 + 0.259 + 0.346 + 0.230$
$= 0.913$

c. X = 2, 3, 4, or 5 people
$P(X) = 0.346 + 0.230 + 0.077 + 0.01$
$= 0.663$

d. X = 0, 1, or 2 people
$P(X) = 0.683$

13.

a. n = 10, p = 0.2, X = 0, 1, 2, 3
$P(X) = 0.107 + 0.268 + 0.302 + 0.201$
$= 0.878$

13. continued

b. n = 10, p = 0.2, X = 3, P(X) = 0.201

c. n = 10, p = 0.2, X = 5, 6, 7, 8, 9, 10
$P(X) = 0.026 + 0.006 + 0.001 + 0 + 0 + 0$
$= 0.033$

14.

a. $\mu = 100(0.75) = 75$
$\sigma^2 = 100(0.75)(0.25) = 18.75$ or 18.8
$\sigma = \sqrt{18.75} = 4.33$ or 4.3

b. $\mu = 300(0.3) = 90$
$\sigma^2 = 300(0.3)(0.7) = 63$
$\sigma = \sqrt{63} = 7.94$ or 7.9

c. $\mu = 20(0.5) = 10$
$\sigma^2 = 20(0.5)(0.5) = 5$
$\sigma = \sqrt{5} = 2.236$ or 2.2

d. $\mu = 10(0.8) = 8$
$\sigma^2 = 10(0.8)(0.2) = 1.6$
$\sigma = \sqrt{1.6} = 1.265$ or 1.3

e. $\mu = 1000(0.1) = 100$
$\sigma^2 = 1000(0.1)(0.9) = 90$
$\sigma = \sqrt{90} = 9.49$ or 9.5

f. $\mu = 500(0.25) = 125$
$\sigma^2 = 500(0.25)(0.75) = 93.75$
$\sigma = \sqrt{93.75} = 9.68$ or 9.7

g. $\mu = 50(\frac{2}{5}) = 20$
$\sigma^2 = 50(\frac{2}{5})(\frac{3}{5}) = 12$
$\sigma = \sqrt{12} = 3.464$ or 3.5

h. $\mu = 36(\frac{1}{6}) = 6$
$\sigma^2 = 36(\frac{1}{6})(\frac{5}{6}) = 5$
$\sigma = \sqrt{5} = 2.236$ or 2.2

15.

n = 800, p = 0.01
$\mu = 800(0.01) = 8$
$\sigma^2 = 800(0.01)(0.99) = 7.9$
$\sigma = \sqrt{7.92} = 2.8$

17.

n = 500, p = 0.02
$\mu = 500(0.02) = 10$
$\sigma^2 = 500(0.02)(0.98) = 9.8$
$\sigma = \sqrt{9.8} = 3.1$

19.

$n = 1000$, $p = 0.21$

$\mu = 1000(0.21) = 210$

$\sigma^2 = 1000(0.21)(0.79) = 165.9$

$\sigma = \sqrt{165.9} = 12.9$

21.

$n = 18$, $p = 0.25$, $X = 5$

$P(X) = \frac{18!}{13! \, 5!}(0.25)^5(0.75)^{13} = 0.199$

23.

$n = 10$, $p = \frac{1}{3}$, $X = 0, 1, 2, 3$

$P(X) = \frac{10!}{10! \, 0!}(\frac{1}{3})^0(\frac{2}{3})^{10} + \frac{10!}{9! \, 1!}(\frac{1}{3})^1(\frac{2}{3})^9$

$+ \frac{10!}{8! \, 2!}(\frac{1}{3})^2(\frac{2}{3})^8 + \frac{10!}{7! \, 3!}(\frac{1}{3})^3(\frac{2}{3})^7 = 0.559$

25.

$n = 5$, $p = 0.13$, $X = 3, 4, 5$

$P(X) = \frac{5!}{2! \, 3!}(0.13)^3(0.87)^2 +$

$\frac{5!}{1! \, 4!}(0.13)^4(0.87)^1 + \frac{5!}{0! \, 5!}(0.13)^5(0.87)^0$

$= 0.018$

27.

$n = 12$, $p = 0.86$, $X = 10, 11, 12$

$P(X) = \frac{12!}{2! \, 10!}(0.86)^{10}(0.14)^2 +$

$\frac{12!}{1! \, 11!}(0.86)^{11}(0.14)^1 + \frac{12!}{0! \, 12!}(0.86)^{12}(0.14)^0$

$= 0.77$

Yes. The probability is high, 77%.

29.

$n = 5$, $p = 0.2$, $X = 0, 1, 2, 3, 4, 5$

X	0	1	2	3	4	5
P(X)	0.328	0.410	0.205	0.051	0.006	0

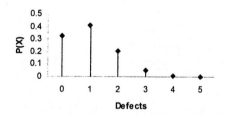

EXERCISE SET 5-5

1.

a. $P(M) = \frac{6!}{3! \, 2! \, 1!}(0.5)^3(0.3)^2(0.2)^1 = 0.135$

b. $P(M) = \frac{5!}{1! \, 2! \, 2!}(0.3)^1(0.6)^2(0.1)^2 = 0.0324$

c. $P(M) = \frac{4!}{1! \, 1! \, 2!}(0.8)^1(0.1)^1(0.1)^2 = 0.0096$

d. $P(M) = \frac{3!}{1! \, 1! \, 1!}(0.5)^1(0.3)^1(0.2)^1 = 0.18$

1. continued

e. $P(M) = \frac{5!}{1! \, 3! \, 1!}(0.7)^1(0.2)^3(0.1)^1 = 0.0112$

3.

$P(M) = \frac{8!}{3! \, 2! \, 3!}(0.25)^3(0.40)^2(0.35)^3 = 0.06$

5.

$P(M) = \frac{4!}{2! \, 1! \, 1!}(\frac{1}{6})^2(\frac{1}{6})^1(\frac{1}{6})^1 = \frac{1}{108}$

7.

a. $P(5; 4) = 0.1563$

b. $P(2; 4) = 0.1465$

c. $P(6; 3) = 0.0504$

d. $P(10; 7) = 0.071$

e. $P(9; 8) = 0.1241$

9.

$p = \frac{1}{20,000} = 0.00005$

$\lambda = n \cdot p = 80,000(0.00005) = 4$

a. $P(0; 4) = 0.0183$

b. $P(1; 4) = 0.0733$

c. $P(2; 4) = 0.1465$

d. $P(3 \text{ or more}; 4) = 1 - [P(0; 4) + P(1; 4)$

$+ P(2; 4)]$

$= 1 - (0.0183 + 0.0733 + 0.1465)$

$= 0.7619$

11.

$p = \frac{5}{1000} = \frac{1}{200}$

$\lambda = n \cdot p = (250) \cdot (\frac{1}{200}) = 1.25$

$P(\text{at least 2 orders}) =$

$1 - [P(0 \text{ orders}) + P(1 \text{ order})]$

$= 1 - [\frac{e^{-1.25}(1.25)^0}{0!} + \frac{e^{-1.25}(1.25)^1}{1!}]$

$= 1 - (0.2865 + 0.3581) = 0.3554$

13.

$\lambda = \frac{1}{1000} \cdot 3000 = 3$

$P(1 \text{ or more}; 3) = 0.1494 + 0.2240 +$

$0.2240 + 0.1680 + 0.1008 + 0.0504 +$

$0.0216 + 0.0081 + 0.0027 + 0.0008 +$

$0.0002 + 0.001 = 0.9502$

15.

$P(5; 4) = 0.1563$

17.

$a = 9$ prefer hoods

$b = 9$ prefer hats

$X = 3$, $n = 6$

$P(A) = \frac{_9C_3 \cdot _9C_3}{_{18}C_6} = \frac{84}{221} = 0.38$

19.
$a = 15, b = 9, n = 4, X = 4$

$$P(A) = \frac{{}_{15}C_4 \cdot {}_9C_0}{{}_{24}C_4} = \frac{65}{506} = 0.13$$

21.
P(at least 1 defective) = 1 − P(0 defectives)
$a = 6, b = 18, n = 3, X = 0$

$$P(0) = \frac{{}_6C_0 \cdot {}_{18}C_3}{{}_{24}C_3} = \frac{102}{253} = 0.403$$
P(at least 1 defective) = 1 − 0.403 = 0.597

REVIEW EXERCISES - CHAPTER FIVE

1.
No, the sum of the probabilities is greater than one.

3.
No, the sum of the probabilities is greater than one.

5.

X	0	1	2	3	4
P(X)	0.05	0.30	0.45	0.12	0.08

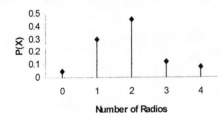

7.

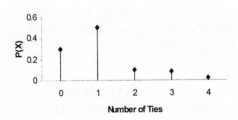

9.
$\mu = \sum X \cdot P(X) = 8(0.15) + 9(0.25) + 10(0.29) + 11(0.19) + 12(0.12) = 9.88$ or 9.9

9. continued
$\sigma^2 = \sum X^2 \cdot P(X) - \mu^2 = [8^2(0.15) + 9^2(0.25) + 10^2(0.29) + 11^2(0.19) + 12^2(0.12)] - 9.88^2 = 1.5056$ or 1.5

$\sigma = \sqrt{1.5056} = 1.23$ or 1.2

X	P(X)	X · P(X)	X² · P(X)
8	0.15	1.20	9.60
9	0.25	2.25	20.25
10	0.29	2.90	29.00
11	0.19	2.09	22.99
12	0.12	1.44	17.28
$\mu =$		9.88	99.12

11.
$\mu = \sum X \cdot P(X) = 22(0.08) + 23(0.19) + 24(0.36) + 25(0.25) + 26(0.07) + 27(0.05)$
$= 24.19$ or 24.2

$\sigma^2 = \sum X^2 \cdot P(X) - \mu^2 = [22^2(0.08) + 23^2(0.19) + 24^2(0.36) + 25^2(0.25) + 26^2(0.07) + 27^2(0.05)] - 24.19^2 = 1.4539$ or 1.5

$\sigma = \sqrt{1.4539} = 1.206$ or 1.2

X	P(X)	X · P(X)	X² · P(X)
22	0.08	1.76	38.72
23	0.19	4.37	100.51
24	0.36	8.64	207.36
25	0.25	6.25	156.25
26	0.07	1.82	47.32
27	0.05	1.35	36.45
$\mu =$		24.19	586.61

13.
$\mu = \sum X \cdot P(X)$
$= \frac{1}{2}(\$1.00) + \frac{18}{52}(\$5.00) + \frac{6}{52}(\$10.00) + \frac{2}{52}(\$100.00) = \$7.23$

To break even, a person should bet \$7.23.

15.
a. 0.122
b. 1 − 0.002 + 0.009 = 0.989
c. 0.002 + 0.009 + 0.032 = 0.043

17.
$\mu = n \cdot p = 180(0.75) = 135$
$\sigma^2 = n \cdot p \cdot q = 180(0.75)(0.25) = 33.75$ or 33.8
$\sigma = \sqrt{33.75} = 5.809$ or 5.8

19.

$n = 8, p = 0.25$

$P(X \leq 3) = \frac{8!}{8!\,0!}(0.25)^0(0.75)^8 +$

$\frac{8!}{7!\,1!}(0.25)^1(0.75)^7 + \frac{8!}{6!\,2!}(0.25)^2(0.75)^6 +$

$\frac{8!}{5!\,3!}(0.25)^3(0.75)^5 = 0.8862$ or 0.886

21.

$n = 20, p = 0.75, X = 16$

P(16 have eaten pizza for breakfast) =

$\frac{20!}{4!\,16!}(0.75)^{16}(0.25)^4 = 0.1897$ or 0.190

23.

$P(M) = \frac{20!}{12!\,4!\,3!\,1!}(0.7)^{12}(0.2)^4(0.08)^3(0.02)^1$

$= 0.008$

25.

$P(M) = \frac{10!}{5!\,3!\,2!}(0.50)^5(0.40)^3(0.10)^2 = 0.050$

27.

a. P(6 or more; 6) = 1 − P(5 or less; 6)
 = 1 − (0.0025 + 0.0149 + 0.0446 +
0.0892 + 0.1339 + 0.1606) = 0.5543

b. P(4 or more; 6) = 1 − P(3 or less; 6)
 = 1 − (0.0025 + 0.0149 + 0.0446 +
0.0892) = 0.8488

c. P(5 or less; 6) = P(0; 6) + ... + P(6; 6)
 = 0.4457

29.

$a = 13, b = 39, n = 5, X = 2$

$P(2) = \frac{_{13}C_2 \cdot {}_{39}C_3}{_{52}C_5} = \frac{9{,}139}{33{,}320} = 0.27$

31.

a. $a = 7, b = 5$
$n = 3, X = 2$ men, 1 woman
$P(2) = \frac{_7C_2 \cdot {}_5C_1}{_{12}C_3} = \frac{21}{44}$ or 0.477

b. $a = 7, b = 5$
$n = 3, X = 0$ men, 3 women

$P(0) = \frac{_7C_0 \cdot {}_5C_3}{_{12}C_3} = \frac{1}{22}$ or 0.045

c. $a = 7, b = 5$
$n = 3, X = 1$ man, 2 women

$P(1) = \frac{_7C_1 \cdot {}_5C_2}{_{12}C_3} = \frac{7}{22}$ or 0.318

CHAPTER 5 QUIZ

1. True
2. False, it is a discrete random variable.
3. False, the outcomes must be independent.

4. True
5. chance
6. $\mu = n \cdot p$
7. one
8. c.
9. c.
10. d.
11. No, the sum of the probabilities is greater than one.
12. Yes
13. Yes
14. Yes
15.

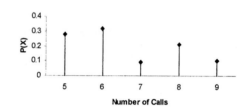

16.

X	0	1	2	3	4
P(X)	0.02	0.30	0.48	0.13	0.07

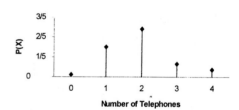

17.

$\mu = 0(0.10) + 1(0.23) + 2(0.31) + 3(0.27)$
$+ 4(0.09) = 2.02$

$\sigma^2 = [0^2(0.10) + 1^2(0.23) + 2^2(0.31) +$
$3^2(0.27) + 4^2(0.09)] - 2.02^2 = 1.3$

$\sigma = \sqrt{1.3} = 1.1$

18.

$\mu = 30(0.05) + 31(0.21) + 32(0.38) +$
$33(0.25) + 34(0.11) = 32.16$ or 32.2

$\sigma^2 = [30^2(0.05) + 31^2(0.21) + 32^2(0.38) +$
$33^2(0.25) + 34^2(0.11)] - 32.16^2 = 1.07$ or
1.1

$\sigma = \sqrt{1.07} = 1.0$

19.

$\mu = 4(\frac{1}{6}) + 5(\frac{1}{6}) + 2(\frac{1}{6}) + 10(\frac{1}{6}) + 3(\frac{1}{6})$
$+ 7(\frac{1}{6}) = 5.17$ or 5.2

20.
$\mu = \$2(\frac{1}{2}) + \$10(\frac{5}{26}) + \$25(\frac{3}{26}) +$
$\$100(\frac{1}{26}) = \$9.65$

21.
n = 20, p = 0.40, X = 5
P(5) = 0.124

22.
n = 20, p = 0.60
a. P(15) = 0.075
b. P(10, 11, ..., 20) = 0.117
c. P(0, 1, 2, 3, 4, 5) = 0.125

23.
n = 300, p = 0.80
$\mu = 300(0.80) = 240$
$\sigma^2 = 300(0.80)(0.20) = 48$
$\sigma = \sqrt{48} = 6.9$

24.
n = 75, p = 0.12
$\mu = 75(0.12) = 9$
$\sigma^2 = 75(0.12)(0.88) = 7.9$
$\sigma = \sqrt{7.9} = 2.8$

25.
$P(M) = \frac{30!}{15!\,8!\,5!\,2!}(0.5)^{15}(0.3)^8(0.15)^5(0.05)^2$

$= 0.0080$

26.
$P(M) = \frac{16!}{9!\,4!\,3!}(0.88)^9(0.08)^4(0.04)^3$

$= 0.0003$

27.
$P(M) = \frac{12!}{5!\,4!\,3!}(0.45)^5(0.35)^4(0.2)^3$

$= 0.061$

28.
$\lambda = 100(0.08) = 8, \ X = 6$
P(6; 8) = 0.122

29.
$\lambda = 8$
a. P(X $\geq$ 8; 8) = 0.1396 + ... + 0.0001
   = 0.5471
b. P(X $\geq$ 3; 8) = 1 − P(0, 1, or 2 calls)
   = 1 − (0.0003 + 0.0027 + 0.0107)
   = 1 − 0.0137 = 0.9863
c. P(X $\leq$ 7; 8) = 0.0003 + ... + 0.1396
   = 0.4529

30.
a = 12, b = 36, n = 6, X = 3

$P(A) = \frac{_{12}C_3 \cdot {_{36}C_3}}{_{48}C_6} = \frac{\frac{12!}{9!\,3!} \cdot \frac{36!}{33!\,3!}}{\frac{48!}{42!\,6!}} = 0.128$

31.
a. $\frac{_6C_3 \cdot {_8C_1}}{_{14}C_4} = \frac{\frac{6!}{3!\,3!} \cdot \frac{8!}{7!\,1!}}{\frac{14!}{10!\,4!}} = 0.16$

b. $\frac{_6C_2 \cdot {_8C_2}}{_{14}C_4} = \frac{\frac{6!}{4!\,2!} \cdot \frac{8!}{6!\,2!}}{\frac{14!}{10!\,4!}} = 0.42$

c. $\frac{_6C_0 \cdot {_8C_4}}{_{14}C_4} = \frac{\frac{6!}{6!\,0!} \cdot \frac{8!}{4!\,4!}}{\frac{14!}{10!\,4!}} = 0.07$

Note: Graphs are not to scale and are intended to convey a general idea.

Answers are generated using Table E. Answers generated using the TI-83 or computer programs will vary slightly.

EXERCISE SET 6-3

1.
The characteristics of the normal distribution are:
1. It is bell-shaped.
2. It is symmetric about the mean.
3. The mean, median, and mode are equal.
4. It is continuous.
5. It never touches the X-axis.
6. The area under the curve is equal to one.
7. It is unimodal.

3.
One or 100%.

5.
68%, 95%, 99.7%

7.
The area is found by looking up z = 0.56 in Table E as shown in Block 1 of Procedure Table 6. Area = 0.2123

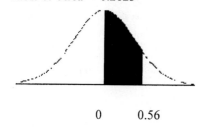

0   0.56

9.
The area is found by looking up z = 2.07 in Table E as shown in Block 1 of Procedure Table 6. Area = 0.4808

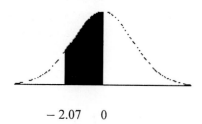

− 2.07   0

11.
The area is found by looking up z = 0.23 in Table E and subtracting it from 0.5 as shown in Block 2 of Procedure Table 6.
0.5 − 0.0910 = 0.4090

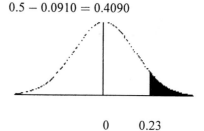

0   0.23

13.
The area is found by looking up z = 1.43 in Table E and subtracting it from 0.5 as shown in Block 2 of Procedure Table 6.
0.5 − 0.4236 = 0.0764

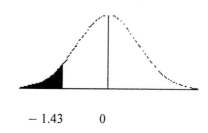

− 1.43   0

15.
The area is found by looking up the values 0.79 and 1.28 in Table E and subtracting the areas as shown in Block 3 of Procedure Table 6.   0.3997 − 0.2852 = 0.1145

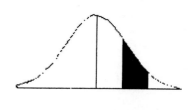

0   0.79   1.28

17.
The area is found by looking up the values 1.56 and 1.83 in Table E and subtracting the areas as shown in Block 3 of Procedure Table 6.   0.4664 − 0.4406 = 0.0258

17. continued

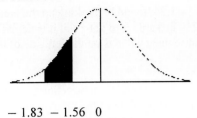

$-1.83 \quad -1.56 \quad 0$

19.
The area is found by looking up the values 2.47 and 1.03 in Table E and adding them together as shown in Block 4 of Procedure Table 6. $0.3485 + 0.4932 = 0.8417$

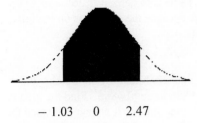

$-1.03 \quad 0 \quad 2.47$

21.
The area is found by looking up $z = 2.11$ in Table E, then adding the area to 0.5 as shown in Block 5 of Procedure Table 6.
$0.5 + 0.4826 = 0.9826$

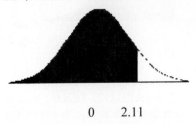

$0 \quad 2.11$

23.
The area is found by looking up $z = 0.18$ in Table E and adding it to 0.5 as shown in Block 6 of Procedure Table 6.
$0.5 + 0.0714 = 0.5714$

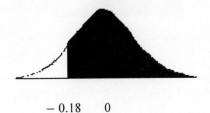

$-0.18 \quad 0$

25.
The area is found by looking up the values 1.92 and -0.44 in Table E, subtracting both areas from 0.5, and adding them

25. continued
together as shown in Block 7 of Procedure Table 6.
$0.5 - 0.4726 = 0.0274$
$0.5 - 0.1700 = 0.3300$
$0.0274 + 0.3300 = 0.3574$

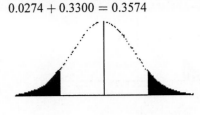

$-0.44 \quad 0 \quad 1.92$

27.
The area is found by looking up $z = 0.67$ in Table E as shown in Block 1 of Procedure Table 6. Area $= 0.2486$

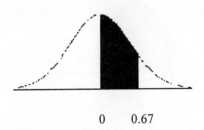

$0 \quad 0.67$

29.
The area is found by looking up $z = 1.57$ in Table E as shown in Block 1 of Procedure Table 6. Area $= 0.4418$

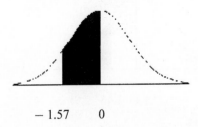

$-1.57 \quad 0$

31.
The area is found by looking up $z = 2.83$ in Table E then subtracting the area from 0.5 as shown in Block 2 of Procedure Table 6.
$0.5 - 0.4977 = 0.0023$

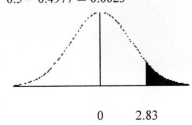

$0 \quad 2.83$

33.
The area is found by looking up z = 1.51 in Table E then subtracting the area from 0.5 as shown in Block 2 of Procedure Table 6.
0.5 − 0.4345 = 0.0655

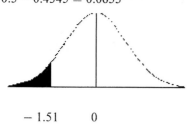

− 1.51   0

35.
The area is found by looking the values 2.46 and 1.74 in Table E and adding the areas together as shown in Block 4 of Procedure Table 6. 0.4931 + 0.4591 = 0.9522

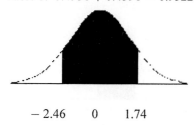

− 2.46   0   1.74

37.
The area is found by looking up the values 1.46 and 2.97 in Table E and subtracting the areas as shown in Block 3 of Procedure Table 6. 0.4985 − 0.4279 = 0.0706

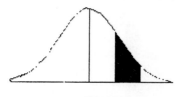

0   1.46   2.97

39.
The area is found by looking up z = 1.42 in Table E and adding 0.5 to it as shown in Block 5 of Procedure Table 6.
0.5 + 0.4222 = 0.9222

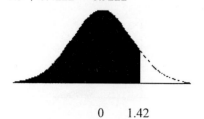

0   1.42

41.
z = − 1.94, found by looking up the area 0.4738 in Table E to get 1.94; it is negative because the z value is on the left side of 0.

43.
z = − 2.13, found by subtracting 0.0166 from 0.5 to get 0.4834 then looking up the area to get z = 2.13; it is negative because the z value is on the left side of 0.

45.
z = − 1.26, found by subtracting 0.5 from 0.8962 to get 0.3962, then looking up the area in Table E to get z = 1.26; it is negative because the z value is on the left side of 0.

47.
a. z = − 2.28, found by subtracting 0.5 from 0.9886 to get 0.4886. Find the area in Table E, then find z. It is negative since the z value falls to the left of 0.

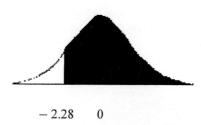

− 2.28   0

b. z = − 0.92, found by subtracting 0.5 from 0.8212 to get 0.3212. Find the area in Table E, then find z. It is negative since the z value falls to the left of 0.

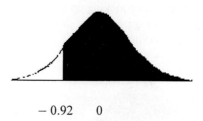

− 0.92   0

c. z = − 0.27, found by subtracting 0.5 from 0.6064 to get 0.1064. Find the area in Table E, then find z. It is negative since the z value falls to the left of 0.

48

47c. continued

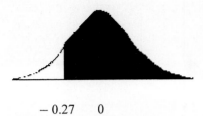

−0.27   0

49.

a. z = ± 1.96, found by:
0.05 ÷ 2 = 0.025 is the area in each tail.
0.5 − 0.025 = 0.4750 is the area needed to determine z.

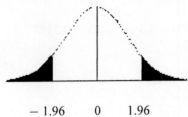

−1.96   0   1.96

b. z = ± 1.65, found by:
0.10 ÷ 2 = 0.05 is the area in each tail.
0.5 − 0.05 = 0.4500 is the area needed to determine z.

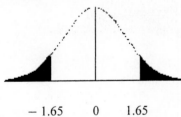

−1.65   0   1.65

c. z = ± 2.58, found by:
0.01 ÷ 2 = 0.005 is the area in each tail.
0.5 − 0.005 = 0.4950 is the area needed to determine z.

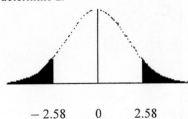

−2.58   0   2.58

51.
$P(-1 < z < 1) = 2(0.3413) = 0.6826$

$P(-2 < z < 2) = 2(0.4772) = 0.9544$

$P(-3 < x < 3) = 2(0.4987) = 0.9974$

51. continued
They are very close.

53.
For z = − 1.2, area = 0.3849
0.8671 − 0.3849 = 0.4822
For area = 0.4822, z = 2.10
Thus, $P(-1.2 < z < 2.10) = 0.8671$

55.
For z = − 0.5, area = 0.1915
0.2345 − 0.1915 = 0.043
For area = 0.043, z = 0.11
Thus, $P(-0.5 < z < 0.11) = 0.2345$

For z = − 0.5, area = 0.1915
0.2345 + 0.1915 = 0.4260
For area = 0.426, z = − 1.45
Thus, $P(-1.45 < z < -0.5) = 0.2345$

57.
$$y = \frac{e^{-\frac{(X-0)^2}{2(1)^2}}}{1\sqrt{2\pi}} = \frac{e^{-\frac{X^2}{2}}}{\sqrt{2\pi}}$$

EXERCISE SET 6-4

1.
$z = \frac{\$3.00 - \$5.39}{\$0.79} = -3.03$
area = 0.4988
$P(z < -3.03) = 0.5 - 0.4988 = 0.0012$ or 0.12%

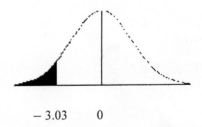

−3.03   0

3.
$z = \frac{X - \mu}{\sigma}$

a. $z = \frac{700,000 - 618,319}{50,200} = 1.63$
area = 0.4484

$P(z > 1.63) = 0.5 - 0.4484 = 0.0516$ or 5.16%

3. continued

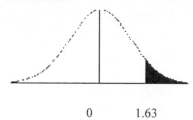

0   1.63

b. $z = \frac{500,000-618,319}{50,200} = -2.36$
area = 0.4909

$z = \frac{600,000-618,319}{50,200} = -0.36$
area = 0.1406

$P(-2.36 < z < -0.36) = 0.4909 - 0.1406$
$P = 0.3503$ or 35.03%

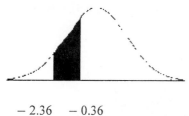

-2.36 -0.36

5.
$z = \frac{X-\mu}{\sigma}$

a. $z = \frac{200-225}{10} = -2.5$
area = 0.4938

$z = \frac{220-225}{10} = -0.5$
area = 0.1915

$P(-2.5 < z < -0.5) =$
$0.4938 - 0.1915 = 0.3023$ or 30.23%

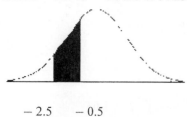

-2.5 -0.5

b. $z = -2.5$
area = 0.4938

$P(z < -2.5) = 0.5 - 0.4938 = 0.0062$ or
0.62%

5b. continued

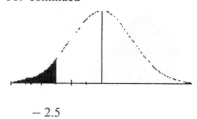

-2.5

7.
$z = \frac{X-\mu}{\sigma}$

a. $z = \frac{\$90,000-\$85,900}{\$11,000} = 0.37$
area = 0.1443

$P(z > 0.37) = 0.5 - 0.1443 = 0.3557$
or 35.57%

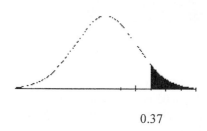

0.37

b. $z = \frac{\$75,000-\$85,900}{\$11,000} = -0.99$
area = 0.3389

$P(z > -0.99) = 0.5 + 0.3389$
$= 0.8389$ or 83.89%

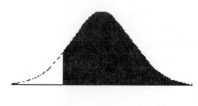

-0.99

9.
$z = \frac{X-\mu}{\sigma}$

a. $z = \frac{180-200}{10} = -2.00$
area = 0.4772

$P(z \geq -2) = 0.5 + 0.4772 = 0.9772$ or
97.72%

9a. continued

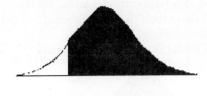

$$-2.00 \quad\quad 0$$

b. $z = \frac{205-200}{10} = 0.5$
area $= 0.1915$

$P(z \leq 0.5) = 0.5 + 0.1915 = 0.6915$ or
69.15%

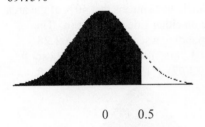

$$0 \quad\quad 0.5$$

11.
$z = \frac{X-\mu}{\sigma}$

a. $z = \frac{1000-3262}{1100} = -2.06$
area $= 0.4803$

$P(z \geq -2.06) = 0.5 + 0.4803 = 0.9803$
or 98.03%

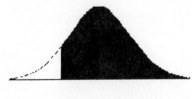

$$-2.06$$

b. $z = \frac{4000-3262}{1100} = 0.67$
area $= 0.2486$

$P(z > 0.67) = 0.5 - 0.2486 = 0.2514$ or
25.14%

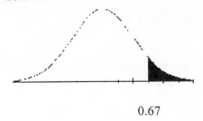

$$0.67$$

c. $z = \frac{3000-3262}{1100} = -0.24$

11c. continued
area $= 0.0948$

$P(-0.24 < z < 0.67) =$
$0.0948 + 0.2486 = 0.3434$ or 34.34%

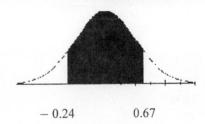

$$-0.24 \quad\quad\quad 0.67$$

13.
a. $z = \frac{280-300}{8} = -2.5$
area $= 0.4938$

$P(z > -2.5) = 0.5 + 0.4938 = 0.9938$ or
99.38%

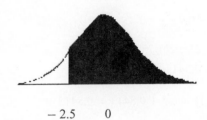

$$-2.5 \quad\quad 0$$

b. $z = \frac{293-300}{8} = -0.88$
area $= 0.3106$

$P(z < -0.88) = 0.5 - 0.3106 = 0.1894$ or
18.94%

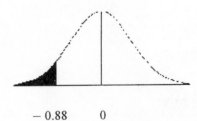

$$-0.88 \quad\quad 0$$

c. $z = \frac{285-300}{8} = -1.88$
area $= 0.4699$

$z = \frac{320-300}{8} = 2.5$
area $= 0.4938$

$P(-1.88 < z < 2.5) =$
$0.4699 + 0.4938 = 0.9637$ or 96.37%

13c. continued

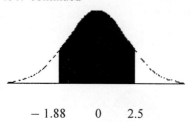

$$-1.88 \quad 0 \quad 2.5$$

15.

a. $z = \frac{130-132}{8} = -0.25$

area $= 0.0987$

$P(z > -0.25) = 0.5 + 0.0987 = 0.5987$ or 59.87%

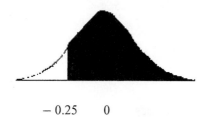

$$-0.25 \quad 0$$

b. $z = \frac{140-132}{8} = 1.00$

area $= 0.3413$

$P(z < 1) = 0.5 + 0.3413 = 0.8413$ or 84.13%

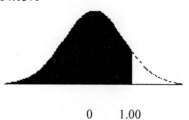

$$0 \quad 1.00$$

c. $z = \frac{131-132}{8} = -0.13$

area $= 0.0517$

$z = \frac{136-132}{8} = 0.50$

area $= 0.1915$

$P(-0.13 < z < 0.50) = 0.0517 + 0.1915$
$= 0.2432$ or 24.32%

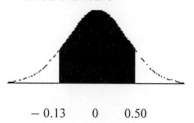

$$-0.13 \quad 0 \quad 0.50$$

17.
The top 75% (area) includes all but the left 25% of the curve. The corresponding z score is -0.67.
$x = -0.67(15) + 100 = 89.95$ points

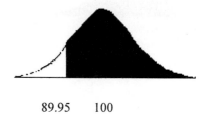

$$89.95 \quad 100$$

19.
The middle 80% means that 40% of the area will be on either side of the mean. The corresponding z scores will be ± 1.28.
$x = -1.28(92) + 1810 = 1694.24$ sq. ft.
$x = 1.28(92) + 1810 = 1927.76$ sq. ft.

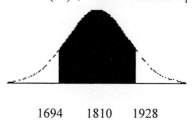

$$1694 \quad 1810 \quad 1928$$

21.
$z = \frac{1200-949}{100} = 2.51$
area $= 0.4940$

$P(z > 2.51) = 0.5 - 0.4940 = 0.006$ or 0.6%

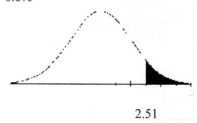

$$2.51$$

For the least expensive 10%, the area is 0.4 on the left side of the curve. Thus,
$z = -1.28$.
$x = -1.28(100) + 949 = \$821$

23.
The middle 60% means that 30% of the area will be on either side of the mean. The corresponding z scores will be ± 0.84.
$x = -0.84(1150) + 8256 = \$7290$
$x = 0.84(1150) + 8256 = \$9222$

23. continued

$7290 $8256 $9222

25.
For the fewest 15%, the area is 0.35 on the left side of the curve. Thus, $z = -1.04$.
$x = -1.04(1.7) + 5.9$
$x = 4.132$ days
For the longest 25%, the area is 0.25 on the right side of the curve. Thus, $z = 0.67$.
$x = 0.67(1.7) + 5.9$
$x = 7.039$ days

27.
The bottom 18% means that 32% of the area is between 0 and $-z$. The corresponding z score will be -0.92.
$x = -0.92(6256) + 24,596 = \$18,840.48$

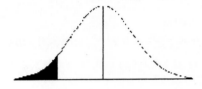

$18,840.48 $24,596

29.
The 10% to be exchanged would be at the left, or bottom, of the curve; therefore, 40% of the area is between 0 and $-z$. The corresponding z score will be -1.28.
$x = -1.28(5) + 25 = 18.6$ months.

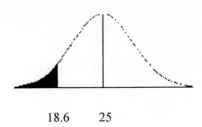

18.6   25

31.
a. $\mu = 120$ $\sigma = 20$
b. $\mu = 15$ $\sigma = 2.5$
c. $\mu = 30$ $\sigma = 5$

33.
There are several mathematical tests that can be used including drawing a histogram and calculating Pearson's index of skewness.

35.
2.68% area in the right tail of the curve means that 47.32% of the area is between 0 and z, corresponding to a z score of 1.93.
$z = \frac{X-\mu}{\sigma}$
$1.93 = \frac{105-100}{\sigma}$
$1.93\sigma = 5$
$\sigma = 2.59$

37.
1.25% of the area in each tail means that 48.75% of the area is between 0 and $\pm z$. The corresponding z scores are ± 2.24.
Then $\mu = \frac{42+48}{2} = 45$ and $X = \mu + z\sigma$.
$48 = 45 + 2.24\sigma$
$\sigma = 1.34$

39.
Histogram:

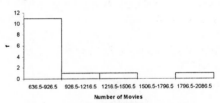

The histogram shows a positive skew.

$PI = \frac{3(970.2-853.5)}{376.5} = 0.93$

$IQR = Q_3 - Q_1 = 910 - 815 = 95$
$1.5(IQR) = 1.5(95) = 142.5$
$Q_1 - 142.5 = 672.5$
$Q_3 + 142.5 = 1052.5$
There are several outliers.

Conclusion: The distribution is not normal.

41.
Histogram:

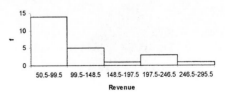

The histogram shows a positive skew.

41. continued

$$PI = \frac{3(115.3 - 92.5)}{66.32} = 1.03$$

$IQR = Q_3 - Q_1 = 154.5 - 67 = 87.5$
$1.5(IQR) = 1.5(87.5) = 131.25$
$Q_1 - 131.25 = -64.25$
$Q_3 + 131.25 = 285.75$
There is one outlier.

Conclusion: The distribution is not normal.

EXERCISE SET 6-5

1.
The distribution is called the sampling distribution of sample means.

3.
The mean of the sample means is equal to the population mean.

5.
The distribution will be approximately normal when sample size is large.

7.
$$z = \frac{\overline{X} - \mu}{\sigma / \sqrt{n}}$$

9.
$$z = \frac{\overline{X} - \mu}{\frac{\sigma}{\sqrt{n}}} = \frac{\$175 - \$186.80}{\frac{\$32}{\sqrt{50}}} = -2.61$$
area $= 0.4955$
$P(z < -2.61) = 0.5 - 0.4955 = 0.0045$ or 0.45%

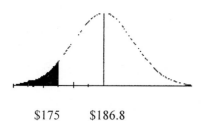

$\$175 \qquad \$186.8$

11.
$$z = \frac{\overline{X} - \mu}{\frac{\sigma}{\sqrt{n}}} = \frac{128.3 - 126}{\frac{15.7}{\sqrt{25}}} = 0.73$$
area $= 0.2673$
$P(z > 0.73) = 0.5 - 0.2673 = 0.2327$ or 23.27%

11. continued

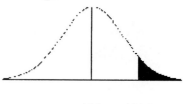

$126 \qquad 128.3$

13.
$$z = \frac{\overline{X} - \mu}{\frac{\sigma}{\sqrt{n}}} = \frac{\$2.00 - \$2.02}{\frac{\$0.08}{\sqrt{40}}} = -1.58$$
area $= 0.4429$
$P(z < -1.58) = 0.5 - 0.4429 = 0.0571$ or 5.71%

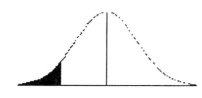

$\$2.00 \qquad \$2.02$

15.
$$z = \frac{\overline{X} - \mu}{\frac{\sigma}{\sqrt{n}}} = \frac{27 - 30}{\frac{5}{\sqrt{22}}} = -2.81$$
area $= 0.4975$
$$z = \frac{\overline{X} - \mu}{\frac{\sigma}{\sqrt{n}}} = \frac{31 - 30}{\frac{5}{\sqrt{22}}} = 0.94 \qquad \text{area} = 0.3264$$
$P(-2.81 < z < 0.94) = 0.4975 + 0.3264$
$= 0.8239$ or 82.39%

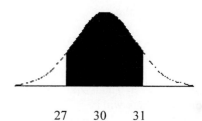

$27 \qquad 30 \qquad 31$

17.
$$z = \frac{\overline{X} - \mu}{\frac{\sigma}{\sqrt{n}}} = \frac{44.2 - 43.6}{\frac{5.1}{\sqrt{50}}} = 0.83$$
area $= 0.2967$
$P(z > 0.83) = 0.5 - 0.2967 = 0.2033$ or 20.33%

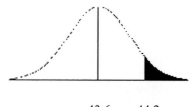

$43.6 \qquad 44.2$

19.

$z = \frac{\overline{X} - \mu}{\frac{\sigma}{\sqrt{n}}} = \frac{1980 - 2000}{\frac{187.5}{\sqrt{50}}} = -0.75$

area = 0.2734

$z = \frac{\overline{X} - \mu}{\frac{\sigma}{\sqrt{n}}} = \frac{1990 - 2000}{\frac{187.5}{\sqrt{50}}} = -0.38$

area = 0.1480

$P(-0.75 < z < -0.38) = 0.2734 - 0.1480 = 0.1254$ or 12.54%

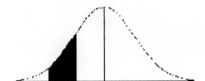

1980 1990 2000

21.

a. $z = \frac{X - \mu}{\sigma} = \frac{43 - 46.2}{8} = -0.4$

area = 0.1554

$P(z < -0.4) = 0.5 - 0.1554 = 0.3446$ or 34.46%

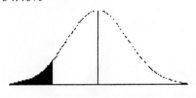

43   46.2

b. $z = \frac{43 - 46.2}{\frac{8}{\sqrt{50}}} = -2.83$ area = 0.4977

$P(z < -2.83) = 0.5 - 0.4977 = 0.0023$ or 0.23%

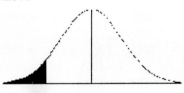

43   46.2

c. Yes, since it is withing one standard deviation of the mean.

d. Very unlikely, since the probability would be less than 1%.

23.

a. $z = \frac{220 - 215}{15} = 0.33$ area = 0.1293

$P(z > 0.33) = 0.5 - 0.1293 = 0.3707$ or 37.07%

23a. continued

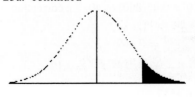

215   220

b. $z = \frac{220 - 215}{\frac{15}{\sqrt{25}}} = 1.67$ area = 0.4525

$P(z > 1.67) = 0.5 - 0.4525 = 0.0475$ or 4.75%

215   220

25.

a. $z_1 = \frac{46 - 48.25}{4.20} = -0.54$ area = 0.2054

$z_2 = \frac{48 - 48.25}{4.20} = -0.06$ area = 0.0239

$P(-0.54 < z < -0.06) = 0.2054 - 0.0239 = 0.1815$ or 18.15%

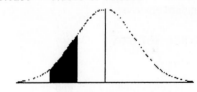

46   48   48.25

b. $z_1 = \frac{46 - 48.25}{\frac{4.20}{\sqrt{20}}} = -2.40$ area = 0.4918

$z_2 = \frac{48 - 48.25}{\frac{4.20}{\sqrt{20}}} = -0.27$ area = 0.1064

$P(-2.40 < z < -0.27) = 0.4918 - 0.1064 = 0.3854$ or 38.54%

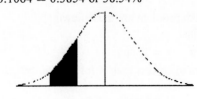

46   48   48.25

c. Means are less variable than individual data.

27.
Since $50 > 0.05(800)$ or 40, the correction factor is necessary.

It is $\sqrt{\frac{800-50}{800-1}} = 0.969$

$z = \frac{\overline{X}-\mu}{\frac{\sigma}{\sqrt{n}} \cdot \sqrt{\frac{N-n}{n-1}}} = \frac{83,500-82,000}{\frac{5000}{\sqrt{50}}(0.969)} = 2.19$

area $= 0.4857$

$P(z > 2.19) = 0.5 - 0.4857 = 0.0143$ or 1.43%

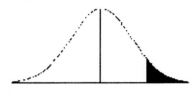

82,000   83,500

29.
$\sigma_x = \frac{\sigma}{\sqrt{n}} = \frac{15}{\sqrt{100}} = 1.5$

$2(1.5) = \frac{15}{\sqrt{n}}$

$3 \cdot \sqrt{n} = 15$

$\sqrt{n} = 5$

$n = 25$, the sample size necessary to double the standard error.

EXERCISE SET 6-6

1.
When p is approximately 0.5, and as n increases, the shape of the binomial distribution becomes similar to the normal distribution. The normal approximation should be used only when $n \cdot p$ and $n \cdot q$ are both greater than or equal to 5. The correction for continuity is necessary because the normal distribution is continuous and the binomial is discrete.

2.
For each problem use the following formulas:

$\mu = np \quad \sigma = \sqrt{npq} \quad z = \frac{\overline{X}-\mu}{\sigma}$

Be sure to correct each X for continuity.

a. $\mu = 0.5(30) = 15$
$\sigma = \sqrt{(0.5)(0.5)(30)} = 2.74$

$z = \frac{17.5-15}{2.74} = 0.91$ area $= 0.3186$

2. continued
$z = \frac{18.5-15}{2.74} = 1.28$ area $= 0.3997$

$P(17.5 < X < 18.5) = 0.3997 - 0.3186$
$= 0.0811 = 8.11\%$

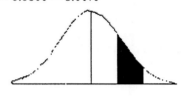

15   17.5 18.5

b. $\mu = 0.8(50) = 40$
$\sigma = \sqrt{(50)(0.8)(0.2)} = 2.83$

$z = \frac{43.5-40}{2.83} = 1.24$ area $= 0.3925$

$z = \frac{44.5-40}{2.83} = 1.59$ area $= 0.4441$

$P(43.5 < X < 44.5) = 0.4441 - 0.3925$
$= 0.0516$ or 5.16%

40   43.5 44.5

c. $\mu = 0.1(100) = 10$
$\sigma = \sqrt{(0.1)(0.9)(100)} = 3$

$z = \frac{11.5-10}{3} = 0.50$ area $= 0.1915$

$z = \frac{12.5-10}{3} = 0.83$ area $= 0.2967$

$P(11.5 < X < 12.5) = 0.2967 - 0.1915$
$= 0.1052$ or 10.52%

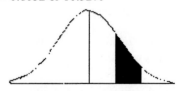

10   11.5 12.5

d. $\mu = 10(0.5) = 5$
$\sigma = \sqrt{(0.5)(0.5)(10)} = 1.58$

$z = \frac{6.5-5}{1.58} = 0.95$ area $= 0.3289$

2d. continued

$P(X \geq 6.5) = 0.5 - 0.3289 = 0.1711$ or 17.11%

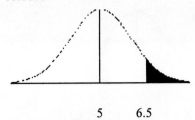

$\qquad$ 5 $\qquad$ 6.5

e. $\mu = 20(0.7) = 14$
$\sigma = \sqrt{(20)(0.7)(0.3)} = 2.05$

$z = \frac{12.5-14}{2.05} = -0.73 \qquad$ area $= 0.2673$

$P(X \leq 12.5) = 0.5 - 0.2673 = 0.2327$ or 23.27%

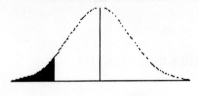

$\qquad$ 12.5 $\qquad$ 14

f. $\mu = 50(0.6) = 30$
$\sigma = \sqrt{(50)(0.6)(0.4)} = 3.46$

$z = \frac{40.5-30}{3.46} = 3.03 \qquad$ area $= 0.4988$

$P(X \leq 40.5) = 0.5 + 0.4988 = 0.9988$ or 99.88%

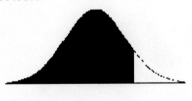

$\qquad$ 30 $\qquad$ 40.5

3.
a. $np = 20(0.50) = 10 \geq 5 \qquad$ Yes
$\quad nq = 20(0.50) = 10 \geq 5$
b. $np = 10(0.60) = 6 \geq 5 \qquad$ No
$\quad nq = 10(0.40) = 4 < 5$
c. $np = 40(0.90) = 36 \geq 5 \qquad$ No
$\quad nq = 40(0.10) = 4 < 5$
d. $np = 50(0.20) = 10 \geq 5 \qquad$ Yes
$\quad nq = 50(0.80) = 40 \geq 5$
e. $np = 30(0.80) = 24 \geq 5 \qquad$ Yes
$\quad nq = 30(0.20) = 6 \geq 5$

3. continued
f. $np = 20(0.85) = 17 \geq 5 \qquad$ No
$\quad nq = 20(0.15) = 3 > 5$

5.
$p = \frac{2}{5} = 0.4 \qquad \mu = 400(0.4) = 160$
$\sigma = \sqrt{(400)(0.4)(0.6)} = 9.8$

$z = \frac{169.5-160}{9.8} = 0.97 \qquad$ area $= 0.3340$

$P(X > 169.5) = 0.5 - 0.3340 = 0.1660$ or 16.6%

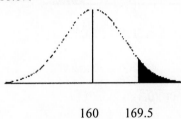

$\qquad$ 160 $\qquad$ 169.5

7.
$\mu = 300(0.509) = 152.7$
$\sigma = \sqrt{(300)(0.509)(0.491)} = 8.66$

$z = \frac{175.5-152.7}{8.66} = 2.63$ area $= 0.4957$

$P(X > 175.5) = 0.5 - 0.4957 = 0.0043$

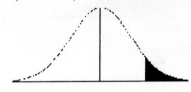

$\qquad$ 152.7 $\qquad$ 175.5

9.
$\mu = 180(0.236) = 42.48$
$\sigma = \sqrt{(180)(0.236)(0.764)} = 5.70$

$z = \frac{50.5-42.48}{5.70} = 1.41 \qquad$ area $= 0.4207$

$P(X > 50.5) = 0.5 - 0.4207 = 0.0793$

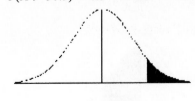

$\qquad$ 42.48 $\qquad$ 50.5

11.

$\mu = 300(0.167) = 50.1$

$\sigma = \sqrt{(300)(0.167)(0.833)} = 6.46$

$z = \frac{50.5 - 50.1}{6.46} = 0.06$ area $= 0.0239$

$P(X > 50.5) = 0.5 - 0.0239 = 0.4761$

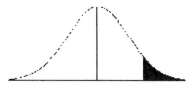

50.1   50.5

13.

$\mu = 400(0.3) = 120$

$\sigma = \sqrt{(400)(0.3)(0.7)} = 9.17$

$z = \frac{99.5 - 120}{9.17} = -2.24$

$P(X > 99.5) = 0.5 + 0.4875 = 0.9875$ or 98.75%

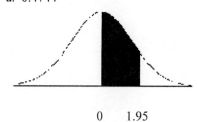

99.5   120

CHAPTER SIX REVIEW EXERCISES

1.

a. 0.4744

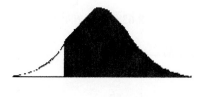

0   1.95

b. 0.1443

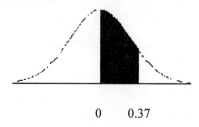

0   0.37

1. continued

c. $0.4656 - 0.4066 = 0.0590$

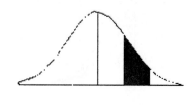

0   1.32   1.82

d. $0.3531 + 0.4798 = 0.8329$

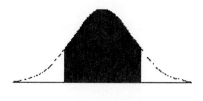

-1.05 0   2.05

e. $0.2019 + 0.0120 = 0.2139$

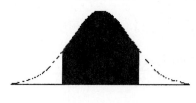

-0.03 0   0.53

f. $0.3643 + 0.4641 = 0.8284$

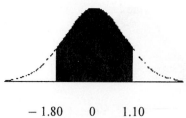

-1.80 0   1.10

g. $0.5 - 0.4767 = 0.0233$

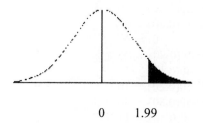

0   1.99

1. continued
h. $0.5 + 0.4131 = 0.9131$

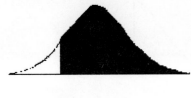

$-1.36 \quad 0$

i. $0.5 - 0.4817 = 0.0183$

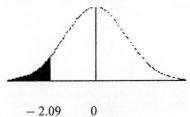

$-2.09 \quad 0$

j. $0.5 + 0.4535 = 0.9535$

$0 \quad 1.68$

3.
a. $z = \frac{10.00 - 8.99}{3.00} = 0.34 \quad$ area $= 0.1331$

$P(z > 0.34) = 0.5 - 0.1331 = 0.3669$

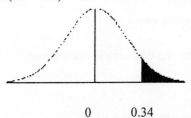

$0 \quad 0.34$

b. $z = \frac{11.00 - 8.99}{3.00} = 0.67 \quad$ area $= 0.2486$

$P(z < 0.67) = 0.5 + 0.2486 = 0.7486$

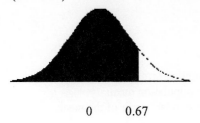

$0 \quad 0.67$

5.
(a) $z = \frac{65 - 63}{8} = 0.25 \quad$ area $= 0.0987$

$P(z > 0.25) = 0.5 - 0.0987 = 0.4013$ or 40.13%

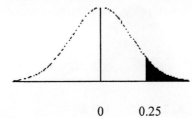

$0 \quad 0.25$

(b) $z = \frac{72 - 63}{8} = 1.13 \quad$ area $= 0.3708$

$P(z > 1.13) = 0.5 - 0.3708 = 0.1292$ or 12.92%

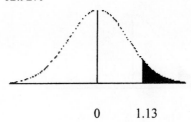

$0 \quad 1.13$

7.
(a) $z = \frac{18 - 19.32}{2.44} = -0.54 \quad$ area $= 0.2054$

$P(z > -0.54) = 0.5 + 0.2054 = 0.7054$

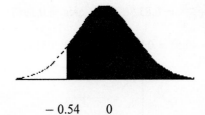

$-0.54 \quad 0$

(b) $z = \frac{18 - 19.32}{\frac{2.44}{\sqrt{5}}} = -1.21 \quad$ area $= 0.3869$

$P(z > -1.21) = 0.5 + 0.3869 = 0.8869$

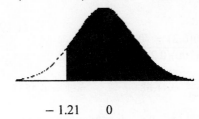

$-1.21 \quad 0$

9.
The middle 40% means that 20% of the area is on either side of the mean. The corresponding z scores are ± 0.52.

9. continued
$X_1 = 100 + (0.52)(15) = 107.8$
$X_2 = 100 + (-0.52)(15) = 92.2$
The scores should be between 92.2 and 107.8.

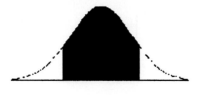

92.2   100   107.8

11.
$z = \frac{11-12.2}{\frac{2.3}{\sqrt{12}}} = -1.81$ area = 0.4649
$P(\overline{X} < 11) = 0.5 - 0.4649 = 0.0351$ or 3.51%

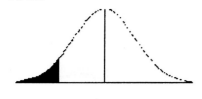

11   12.2

13.
$\mu = 200(0.18) = 36$
$\sigma = \sqrt{(200)(0.18)(0.82)} = 5.43$

$z = \frac{40.5-36}{5.43} = 0.83$ area = 0.2967

$P(X > 40.5) = 0.5 - 0.2967 = 0.2033$ or 20.33%

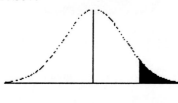

36   40.5

15.
$\mu = 200(0.2) = 40$
$\sigma = \sqrt{(200)(0.2)(0.8)} = 5.66$

$z = \frac{49.5-40}{5.66} = 1.68$ area = 0.4535

$P(X \geq 49.5) = 0.5 - 0.4535 = 0.0465$ or 4.65%

15. continued

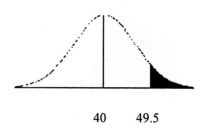

40   49.5

17.
Histogram:

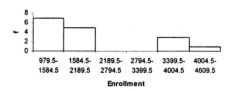

The histogram shows a positive skew.

$PI = \frac{3(2136.1-1755)}{1171.7} = 0.98$

$IQR = Q_3 - Q_1$
$IQR = 2827 - 1320 = 1507$
$1.5(IQR) = 1.5(1507) = 2260.5$
$Q_1 - 2260.5 = -940.5$
$Q_3 + 2260.5 = 5087.5$
There are no outliers.

Conclusion: The distribution is not normal.

CHAPTER 6 QUIZ

1. False, the total area is equal to one.
2. True
3. True
4. True
5. False, the area is positive.
6. False, it applies to means taken from the same population.
7. a.
8. a.
9. b.
10. b.
11. c.
12. 0.5
13. sampling error
14. the population mean
15. the standard error of the mean
16. 5
17. 5%

18. the areas are:

a. 0.4332   f. 0.8079
b. 0.3944   g. 0.0401
c. 0.0344   h. 0..8997
d. 0.1029   i. 0.017
e. 0.2912   j. 0.9131

19. the probabilities are:

a. 0.4846   f. 0.0384
b. 0.4693   g. 0.0089
c. 0.9334   h. 0.9582
d. 0.0188   i. 0.9788
e. 0.7461   j. 0.8461

20. the probabilities are:

a. 0.0531   c. 0.1056
b. 0.1056   d. 0.0994

21. the probabilities are:

a. 0.0668   c. 0.4649
b. 0.0228   d. 0.0934

22. the probabilities are:

a. 0.4525   c. 0.3707
b. 0.3707   d. 0.019

23. the probabilities are:

a. 0.0013   c. 0.0081
b. 0.5   d. 0.5511

24. the probabilities are:

a. 0.0037   c. 0.5
b. 0.0228   d. 0.3232

25. 8.804 cm
26. The lowest acceptable score is 121.24.
27. 0.015
28. 0.9738
29. 0.2296
30. 0.0630
31. 0.8577
32. 0.0495
33. The distribution is not normal.
34. The distribution is normal.

Note: Answers may vary due to rounding.

EXERCISE SET 7-2

1.
A point estimate of a parameter specifies a specific value such as $\mu = 87$, whereas an interval estimate specifies a range of values for the parameter such as $84 < \mu < 90$. The advantage of an interval estimate is that a specific confidence level (say 95%) can be selected, and one can be 95% confident that the parameter being estimated lies in the interval.

3.
The maximum error of estimate is the likely range of values to the right or left of the statistic in which may contain the parameter.

5.
A good estimator should be unbiased, consistent, and relatively efficient.

7.
To determine sample size, the maximum error of estimate and the degree of confidence must be specified and the population standard deviation must be known.

9.
a. 2.58   d. 1.65
b. 2.33   e. 1.88
c. 1.96

11.
a. $\overline{X} - z_{\frac{\alpha}{2}}\left(\frac{s}{\sqrt{n}}\right) < \mu < \overline{X} + z_{\frac{\alpha}{2}}\left(\frac{s}{\sqrt{n}}\right)$

$82 - (1.96)\left(\frac{15}{\sqrt{35}}\right) < \mu < 82 + (1.96)\left(\frac{15}{\sqrt{35}}\right)$

$82 - 4.97 < \mu < 82 + 4.97$

$77 < \mu < 87$

b. $82 - (2.58)\left(\frac{15}{\sqrt{35}}\right) < \mu < 82 + (2.58)\left(\frac{15}{\sqrt{35}}\right)$

$82 - 6.54 < \mu < 82 + 6.54$

$75 < \mu < 89$

c. The 99% confidence interval is larger because the confidence level is larger.

13.
a. $\overline{X} - z_{\frac{\alpha}{2}}\left(\frac{\sigma}{\sqrt{n}}\right) < \mu < \overline{X} + z_{\frac{\alpha}{2}}\left(\frac{\sigma}{\sqrt{n}}\right)$

$12.6 - 1.65\left(\frac{2.5}{\sqrt{40}}\right) < \mu < 12.6 + 1.65\left(\frac{2.5}{\sqrt{40}}\right)$

$12.6 - 0.652 < \mu < 12.6 + 0.652$

13a. continued
$11.9 < \mu < 13.3$

b. It would be highly unlikely since this is far larger than 13.3 minutes.

15.
$\overline{X} - z_{\frac{\alpha}{2}}\left(\frac{s}{\sqrt{n}}\right) < \mu < \overline{X} + z_{\frac{\alpha}{2}}\left(\frac{s}{\sqrt{n}}\right)$

$\$150{,}000 - 1.96\left(\frac{15{,}000}{\sqrt{35}}\right) < \mu <$
$\qquad \$150{,}000 + 1.96\left(\frac{15{,}000}{\sqrt{35}}\right)$

$\$150{,}000 - 4969.51 < \mu <$
$\qquad \$150{,}000 + 4969.51$

$\$145{,}030 < \mu < \$154{,}970$

17.
$\overline{X} = 3222.4 \quad s = 3480.1$

$3222.43 - 1.645\left(\frac{3480.11}{\sqrt{40}}\right) < \mu < 3222.43 +$
$\qquad 1.65\left(\frac{3480.11}{\sqrt{40}}\right)$

$2317.3 < \mu < 4127.6$

19.
$\overline{X} - z_{\frac{\alpha}{2}}\left(\frac{s}{\sqrt{n}}\right) < \mu < \overline{X} + z_{\frac{\alpha}{2}}\left(\frac{s}{\sqrt{n}}\right)$

$61.2 - 1.96\left(\frac{7.9}{\sqrt{84}}\right) < \mu < 61.2 + 1.96\left(\frac{7.9}{\sqrt{84}}\right)$

$61.2 - 1.69 < \mu < 61.2 + 1.69$

$59.5 < \mu < 62.9$

21.
$n = \left[\frac{z_{\frac{\alpha}{2}}\,\sigma}{E}\right]^2 = \left[\frac{(1.96)(54.2)}{2}\right]^2$
$\quad = (53.116)^2 = 2821.31 \text{ or } 2822$

23.
$n = \left[\frac{z_{\frac{\alpha}{2}}\,\sigma}{E}\right]^2 = \left[\frac{(1.96)(2.5)}{1}\right]^2$

$\quad = (4.9)^2 = 24.01 \text{ or } 25$

25.
$n = \left[\frac{z_{\frac{\alpha}{2}}\,\sigma}{E}\right]^2 = \left[\frac{(1.65)(8)}{6}\right]^2$

$\quad = (2.2)^2 = 4.84 \text{ or } 5$

EXERCISE SET 7-3

1.
The characteristics of the t-distribution are: It is bell-shaped, symmetrical about the mean, and never touches the x-axis. The mean, median, and mode are equal to 0 and are located at the center of the distribution.

1. continued

The variance is greater than 1. The t-distribution is a family of curves based on degrees of freedom. As sample size increases the t-distribution approaches the normal distribution.

3.

The t-distribution should be used when σ is unknown and n < 30.

4.

a. 2.898 where d. f. $= 17$

b. 2.074 where d. f. $= 22$

c. 2.624 where d. f. $= 14$

d. 1.833 where d. f. $= 9$

e. 2.093 where d. f. $= 19$

5.

$$\overline{X} - t_{\frac{\alpha}{2}}(\tfrac{s}{\sqrt{n}}) < \mu < \overline{X} + t_{\frac{\alpha}{2}}(\tfrac{s}{\sqrt{n}})$$

$$16 - (2.861)(\tfrac{2}{\sqrt{20}}) < \mu < 16 + (2.861)(\tfrac{2}{\sqrt{20}})$$

$$16 - 1.28 < \mu < 16 + 1.28$$

$$15 < \mu < 17$$

7.

$\overline{X} = 33.4 \quad s = 28.7$

$$\overline{X} - t_{\frac{\alpha}{2}}(\tfrac{s}{\sqrt{n}}) < \mu < \overline{X} + t_{\frac{\alpha}{2}}(\tfrac{s}{\sqrt{n}})$$

$$33.4 - 1.746(\tfrac{28.7}{\sqrt{17}}) < \mu < 33.4 + 1.746(\tfrac{28.7}{\sqrt{17}})$$

$$33.4 - 12.2 < \mu < 33.4 + 12.2$$

$$21.2 < \mu < 45.6$$

The point estimate is 33.4 and is close to the actual population mean of 32, which is within the 90% confidence interval. The mean may not be the best estimate since the data value 132 is large and possibly an outlier.

9.

$$\overline{X} - t_{\frac{\alpha}{2}}(\tfrac{s}{\sqrt{n}}) < \mu < \overline{X} + t_{\frac{\alpha}{2}}(\tfrac{s}{\sqrt{n}})$$

$$12,200 - 2.571(\tfrac{200}{\sqrt{6}}) < \mu <$$
$$12,200 + 2.571(\tfrac{200}{\sqrt{6}})$$

$$12,200 - 209.921 < \mu < 12,200 + 209.921$$

$$11,990 < \mu < 12,410$$

11.

$$\overline{X} - t_{\frac{\alpha}{2}}(\tfrac{s}{\sqrt{n}}) < \mu < \overline{X} + t_{\frac{\alpha}{2}}(\tfrac{s}{\sqrt{n}})$$

$$9.3 - 1.703(\tfrac{2}{\sqrt{28}}) < \mu < 9.3 + 1.703(\tfrac{2}{\sqrt{28}})$$

$$9.3 - 0.644 < \mu < 9.3 + 0.644$$

$$8.7 < \mu < 9.9$$

13.

$$\overline{X} - t_{\frac{\alpha}{2}}(\tfrac{s}{\sqrt{n}}) < \mu < \overline{X} + t_{\frac{\alpha}{2}}(\tfrac{s}{\sqrt{n}})$$

$$18.53 - 2.064(\tfrac{3}{\sqrt{25}}) < \mu <$$
$$18.53 + 2.064(\tfrac{3}{\sqrt{25}})$$

$$18.53 - 1.238 < \mu < 18.53 + 1.238$$

$$\$17.29 < \mu < \$19.77$$

15.

$$\overline{X} - t_{\frac{\alpha}{2}}(\tfrac{s}{\sqrt{n}}) < \mu < \overline{X} + t_{\frac{\alpha}{2}}(\tfrac{s}{\sqrt{n}})$$

$$115 - 2.571(\tfrac{6}{\sqrt{6}}) < \mu < 115 + 2.571(\tfrac{6}{\sqrt{6}})$$

$$115 - 6.298 < \mu < 115 + 6.298$$

$$109 < \mu < 121$$

17.

$\overline{X} = 41.6 \quad x = 5.995$

$$\overline{X} - t_{\frac{\alpha}{2}}(\tfrac{s}{\sqrt{n}}) < \mu < \overline{X} + t_{\frac{\alpha}{2}}(\tfrac{s}{\sqrt{n}})$$

$$41.6 - 2.093(\tfrac{5.995}{\sqrt{20}}) < \mu <$$
$$41.6 + 2.093(\tfrac{5.995}{\sqrt{20}})$$

$$41.6 - 2.806 < \mu < 41.6 + 2.806$$

$$38.8 < \mu < 44.4$$

19.

$$\overline{X} - t_{\frac{\alpha}{2}}(\tfrac{s}{\sqrt{n}}) < \mu < \overline{X} + t_{\frac{\alpha}{2}}(\tfrac{s}{\sqrt{n}})$$

$$\$58,219 - 2.052(\tfrac{56}{\sqrt{28}}) < \mu <$$
$$\$58,219 + 2.052(\tfrac{56}{\sqrt{28}})$$

$$\$58,197 < \mu < \$58,241$$

21.

$\overline{X} = 2.175 \quad s = 0.585$

For $\mu > \overline{X} - t_{\frac{\alpha}{2}}(\tfrac{s}{\sqrt{n}})$:

$$\mu > 2.175 - 1.729(\tfrac{0.585}{\sqrt{20}})$$

$$\mu > 2.175 - 0.226$$

Thus, $\mu > \$1.95$ means that one can be 95% confident that the mean revenue is greater than \$1.95.

For $\mu < \overline{X} + t_{\frac{\alpha}{2}}(\tfrac{s}{\sqrt{n}})$:

$$\mu < 2.175 + 1.729(\tfrac{0.585}{\sqrt{20}})$$

$$\mu < 2.175 + 0.226$$

Thus, $\mu < \$2.40$ means that one can be 95% confident that the mean revenue is less than \$2.40.

EXERCISE SET 7-4

1.

a. $\hat{p} = \tfrac{40}{80} = 0.5 \qquad \hat{q} = \tfrac{40}{80} = 0.5$

b. $\hat{p} = \tfrac{90}{200} = 0.45 \qquad \hat{q} = \tfrac{110}{200} = 0.55$

1. continued

c. $\hat{p} = \frac{60}{130} = 0.46$ $\hat{q} = \frac{70}{130} = 0.54$

d. $\hat{p} = \frac{35}{60} = 0.58$ $\hat{q} = \frac{25}{60} = 0.42$

e. $\hat{p} = \frac{43}{95} = 0.45$ $\hat{q} = \frac{52}{95} = 0.55$

2.

For each part, change the percent to a decimal by dividing by 100, and find $\hat{q}$ using $\hat{q} = 1 - \hat{p}$.

a. $\hat{p} = 0.12$ $\hat{q} = 1 - 0.12 = 0.88$
b. $\hat{p} = 0.29$ $\hat{q} = 1 - 0.29 = 0.71$
c. $\hat{p} = 0.65$ $\hat{q} = 1 - 0.65 = 0.35$
d. $\hat{p} = 0.53$ $\hat{q} = 1 - 0.53 = 0.47$
e. $\hat{p} = 0.67$ $\hat{q} = 1 - 0.67 = 0.33$

3.

$\hat{p} = 0.39$ $\hat{q} = 0.61$

$\hat{p} - (z_{\frac{\alpha}{2}})\sqrt{\frac{\hat{p}\hat{q}}{n}} < p < \hat{p} + (z_{\frac{\alpha}{2}})\sqrt{\frac{\hat{p}\hat{q}}{n}}$

$0.39 - (1.96)\sqrt{\frac{(0.39)(0.61)}{1500}} < p <$

$\qquad 0.39 + (1.96)\sqrt{\frac{(0.39)(0.61)}{1500}}$

$0.39 - 0.025 < p < 0.39 + 0.025$
$0.365 < p < 0.415$

5.

$\hat{p} = \frac{X}{n} = \frac{55}{450} = 0.12$
$\hat{q} = 1 - 0.12 = 0.88$

$\hat{p} - (z_{\frac{\alpha}{2}})\sqrt{\frac{\hat{p}\hat{q}}{n}} < p < \hat{p} + (z_{\frac{\alpha}{2}})\sqrt{\frac{\hat{p}\hat{q}}{n}}$

$0.12 - 1.96\sqrt{\frac{(0.12)(0.88)}{450}} < p < 0.12 + 1.96\sqrt{\frac{(0.12)(0.88)}{450}}$

$0.12 - 0.03 < p < 0.12 + 0.03$
0.09 or $9\% < p < 0.15$ or 15%
11% is contained in the confidence interval.

7.

$\hat{p} = 0.84$ $\hat{q} = 0.16$

$\hat{p} - (z_{\frac{\alpha}{2}})\sqrt{\frac{\hat{p}\hat{q}}{n}} < p < \hat{p} + (z_{\frac{\alpha}{2}})\sqrt{\frac{\hat{p}\hat{q}}{n}}$

$0.84 - 1.65\sqrt{\frac{(0.84)(0.16)}{200}} < p <$

$\qquad 0.84 + 1.65\sqrt{\frac{(0.84)(0.16)}{200}}$

$0.84 - 0.043 < p < 0.84 + 0.043$
$0.797 < p < 0.883$

9.

$\hat{p} = 0.23$ $\hat{q} = 0.77$

$\hat{p} - (z_{\frac{\alpha}{2}})\sqrt{\frac{\hat{p}\hat{q}}{n}} < p < \hat{p} + (z_{\frac{\alpha}{2}})\sqrt{\frac{\hat{p}\hat{q}}{n}}$

$0.23 - 2.58\sqrt{\frac{(0.23)(0.77)}{200}} < p <$

$\qquad 0.23 + 2.58\sqrt{\frac{(0.23)(0.77)}{200}}$

9. continued

$0.23 - 0.077 < p < 0.23 + 0.077$
$0.153 < p < 0.307$
The statement that one in five or 20% of 13 to 14 year olds is a sometime smoker is within the interval.

11.

$\hat{p} = \frac{40}{90} = 0.44$ $\hat{q} = \frac{50}{90} = 0.56$

$\hat{p} - (z_{\frac{\alpha}{2}})\sqrt{\frac{\hat{p}\hat{q}}{n}} < p < \hat{p} + (z_{\frac{\alpha}{2}})\sqrt{\frac{\hat{p}\hat{q}}{n}}$

$0.44 - 1.96\sqrt{\frac{(0.44)(0.56)}{90}} < p <$

$\qquad 0.44 + 1.96\sqrt{\frac{(0.44)(0.56)}{90}}$

$0.44 - 0.103 < p < 0.44 + 0.103$
$0.337 < p < 0.543$

13.

$\hat{p} = 0.44975$ $\hat{q} = 0.55025$

$\hat{p} - (z_{\frac{\alpha}{2}})\sqrt{\frac{\hat{p}\hat{q}}{n}} < p < \hat{p} + (z_{\frac{\alpha}{2}})\sqrt{\frac{\hat{p}\hat{q}}{n}}$

$0.44975 - 1.96\sqrt{\frac{(0.44975)(0.55025)}{1005}} < p <$

$\qquad 0.44975 + 1.96\sqrt{\frac{(0.44975)(0.55025)}{1005}}$

$0.44975 - 0.03076 < p <$

$\qquad 0.44975 + 0.03076$

$0.419 < p < 0.481$

15.

a. $\hat{p} = 0.25$ $\hat{q} = 0.75$

$n = \hat{p}\,\hat{q}\left[\frac{z_{\frac{\alpha}{2}}}{E}\right]^2 = (0.25)(0.75)\left[\frac{2.58}{0.02}\right]^2$

$\quad = 3120.1875$ or 3121

b. $\hat{p} = 0.5$ $\hat{q} = 0.5$

$n = \hat{p}\,\hat{q}\left[\frac{z_{\frac{\alpha}{2}}}{E}\right]^2 = (0.5)(0.5)\left[\frac{2.58}{0.02}\right]^2$

$n = 4160.25$ or 4161

17.

a. $\hat{p} = \frac{30}{300} = 0.1$ $\hat{q} = \frac{270}{300} = 0.9$

$n = \hat{p}\,\hat{q}\left[\frac{z_{\frac{\alpha}{2}}}{E}\right]^2 = (0.1)(0.9)\left[\frac{1.65}{0.05}\right]^2$

$\quad = 98.01$ or 99

b. $\hat{p} = 0.5$ $\hat{q} = 0.5$

$n = \hat{p}\,\hat{q}\left[\frac{z_{\frac{\alpha}{2}}}{E}\right]^2 = (0.5)(0.5)\left[\frac{1.65}{0.05}\right]^2$

$n = 272.25$ or 273

19.

$\hat{p} = 0.5 \qquad \hat{q} = 0.5$

$n = \hat{p}\,\hat{q}\left[\frac{z_{\frac{\alpha}{2}}}{E}\right]^2$

$n = (0.5)(0.5)\left[\frac{1.96}{0.03}\right]^2$

$n = 1067.11$ or 1068

EXERCISE SET 7-5

1.

χ^2

3.

	χ^2_{left}	χ^2_{right}
a.	6.262	27.488
b.	0.711	9.488
c.	8.643	42.796
d.	15.308	44.461
e.	5.892	22.362

5.

$\frac{(n-1)s^2}{\chi^2_{\text{right}}} < \sigma^2 < \frac{(n-1)s^2}{\chi^2_{\text{left}}}$

$\frac{26(6.8)^2}{38.885} < \sigma^2 < \frac{26(6.8)^2}{15.379}$

$30.9 < \sigma^2 < 78.2$
$5.6 < \sigma < 8.8$

7.

$s^2 = 0.80997$ or 0.81

$\frac{(n-1)s^2}{\chi^2_{\text{right}}} < \sigma^2 < \frac{(n-1)s^2}{\chi^2_{\text{left}}}$

$\frac{19(0.81)}{38.582} < \sigma^2 < \frac{19(0.81)}{6.844}$

$0.40 < \sigma^2 < 2.25$
$0.63 < \sigma < 1.50$

9.

$\frac{(n-1)s^2}{\chi^2_{\text{right}}} < \sigma^2 < \frac{(n-1)s^2}{\chi^2_{\text{left}}}$

$\frac{19(19.1913)^2}{30.144} < \sigma^2 < \frac{19(19.1913)^2}{10.117}$

$232.1 < \sigma^2 < 691.6$
$15.2 < \sigma < 26.3$

11.

$\frac{(n-1)s^2}{\chi^2_{\text{right}}} < \sigma^2 < \frac{(n-1)s^2}{\chi^2_{\text{left}}}$

$\frac{27(5.2)^2}{43.194} < \sigma^2 < \frac{27(5.2)^2}{14.573}$

11. continued

$16.9 < \sigma^2 < 50.1$
$4.1 < \sigma < 7.1$

13.

$s - z_{\frac{\alpha}{2}}\left(\frac{s}{\sqrt{2n}}\right) < \sigma < s + z_{\frac{\alpha}{2}}\left(\frac{s}{\sqrt{2n}}\right)$

$18 - 1.96\left(\frac{18}{\sqrt{400}}\right) < \sigma < 18 + 1.96\left(\frac{18}{\sqrt{400}}\right)$

$16.2 < \sigma < 19.8$

REVIEW EXERCISES - CHAPTER 7

1.

$\overline{X} - z_{\frac{\alpha}{2}}\left(\frac{s}{\sqrt{n}}\right) < \mu < \overline{X} + z_{\frac{\alpha}{2}}\left(\frac{s}{\sqrt{n}}\right)$

$2.6 - 1.96\left(\frac{0.4}{\sqrt{36}}\right) < \mu < 2.6 + 1.96\left(\frac{0.4}{\sqrt{36}}\right)$

$2.5 < \mu < 2.7$

3.

$\overline{X} - z_{\frac{\alpha}{2}}\left(\frac{s}{\sqrt{n}}\right) < \mu < \overline{X} + z_{\frac{\alpha}{2}}\left(\frac{s}{\sqrt{n}}\right)$

$7.5 - 1.96\left(\frac{0.8}{\sqrt{1500}}\right) < \mu < 7.5 + 1.96\left(\frac{0.8}{\sqrt{1500}}\right)$

$7.46 < \mu < 7.54$

5.

$\overline{X} - t_{\frac{\alpha}{2}}\left(\frac{s}{\sqrt{n}}\right) < \mu < \overline{X} + t_{\frac{\alpha}{2}}\left(\frac{s}{\sqrt{n}}\right)$

$28 - 2.132\left(\frac{3}{\sqrt{5}}\right) < \mu < 28 + 2.132\left(\frac{3}{\sqrt{5}}\right)$

$25 < \mu < 31$

7.

$n = \left[\frac{z_{\frac{\alpha}{2}}\,\sigma}{E}\right]^2 = \left[\frac{1.65(80)}{25}\right]^2$

$= (5.28)^2 = 27.88$ or 28

9.

$\hat{p} = 0.4 \qquad \hat{q} = 0.6$

$\hat{p} - (z_{\frac{\alpha}{2}})\sqrt{\frac{\hat{p}\hat{q}}{n}} < p < \hat{p} + (z_{\frac{\alpha}{2}})\sqrt{\frac{\hat{p}\hat{q}}{n}}$

$0.4 - 1.65\sqrt{\frac{(0.4)(0.6)}{200}} < p <$
$\qquad 0.4 + 1.65\sqrt{\frac{(0.4)(0.6)}{200}}$

$0.4 - 0.057 < p < 0.4 + 0.057$
$0.343 < p < 0.457$

11.
$$\hat{p} = 0.88 \qquad \hat{q} = 0.12$$

$$n = \hat{p}\,\hat{q}\left[\frac{z_{\frac{\alpha}{2}}}{E}\right]^2 = (0.88)(0.12)\left[\frac{1.65}{0.025}\right]^2$$

$$n = 459.99 \text{ or } 460$$

13.
$$\frac{(n-1)s^2}{\chi^2_{\text{right}}} < \sigma^2 < \frac{(n-1)s^2}{\chi^2_{\text{left}}}$$

$$\frac{(28-1)(0.34)^2}{49.645} < \sigma^2 < \frac{(28-1)(0.34)^2}{11.808}$$

$$0.06287 < \sigma^2 < 0.26433$$

$$0.25 < \sigma < 0.51$$

15.
$$\frac{(n-1)s^2}{\chi^2_{\text{right}}} < \sigma^2 < \frac{(n-1)s^2}{\chi^2_{\text{left}}}$$

$$\frac{(15-1)(8.6)}{23.685} < \sigma^2 < \frac{(15-1)(8.6)}{6.571}$$

$$5.1 < \sigma^2 < 18.3$$

CHAPTER 7 QUIZ

1. True
2. True
3. False, it is consistent if, as sample size increases, the estimator approaches the parameter being estimated.
4. True
5. b.
6. a.
7. b.
8. unbiased, consistent, relatively efficient
9. maximum error of estimate
10. point
11. 90, 95, 99

12. $\overline{X} - z_{\frac{\alpha}{2}}\left(\frac{s}{\sqrt{n}}\right) < \mu < \overline{X} + z_{\frac{\alpha}{2}}\left(\frac{s}{\sqrt{n}}\right)$

$$\$23.45 - 1.65\left(\frac{2.80}{\sqrt{49}}\right) < \mu <$$
$$\$23.45 + 1.65\left(\frac{2.80}{\sqrt{49}}\right)$$

$$\$22.79 < \mu < \$24.11$$

13. $\overline{X} - t_{\frac{\alpha}{2}}\left(\frac{s}{\sqrt{n}}\right) < \mu < \overline{X} + t_{\frac{\alpha}{2}}\left(\frac{s}{\sqrt{n}}\right)$

$$\$44.80 - 2.093\left(\frac{3.53}{\sqrt{20}}\right) < \mu <$$
$$\$44.80 + 2.093\left(\frac{3.53}{\sqrt{20}}\right)$$

13. continued
$$\$43.15 < \mu < \$46.45$$

14. $\overline{X} - z_{\frac{\alpha}{2}}\left(\frac{s}{\sqrt{n}}\right) < \mu < \overline{X} + z_{\frac{\alpha}{2}}\left(\frac{s}{\sqrt{n}}\right)$

$$\$4150 - 2.58\left(\frac{480}{\sqrt{40}}\right) < \mu <$$
$$\$4150 + 2.58\left(\frac{480}{\sqrt{40}}\right)$$

$$\$3954 < \mu < \$4346$$

15. $\overline{X} - t_{\frac{\alpha}{2}}\left(\frac{s}{\sqrt{n}}\right) < \mu < \overline{X} + t_{\frac{\alpha}{2}}\left(\frac{s}{\sqrt{n}}\right)$

$$48.6 - 2.262\left(\frac{4.1}{\sqrt{10}}\right) < \mu < 48.6 + 2.262\left(\frac{4.1}{\sqrt{10}}\right)$$

$$45.7 < \mu < 51.5$$

16. $\overline{X} - t_{\frac{\alpha}{2}}\left(\frac{s}{\sqrt{n}}\right) < \mu < \overline{X} + t_{\frac{\alpha}{2}}\left(\frac{s}{\sqrt{n}}\right)$

$$438 - 3.499\left(\frac{16}{\sqrt{8}}\right) < \mu < 438 + 3.499\left(\frac{16}{\sqrt{8}}\right)$$

$$418 < \mu < 458$$

17. $\overline{X} - t_{\frac{\alpha}{2}}\left(\frac{s}{\sqrt{n}}\right) < \mu < \overline{X} + t_{\frac{\alpha}{2}}\left(\frac{s}{\sqrt{n}}\right)$

$$31 - 2.353\left(\frac{4}{\sqrt{4}}\right) < \mu < 31 + 2.353\left(\frac{4}{\sqrt{4}}\right)$$

$$26 < \mu < 36$$

18. $n = \left[\frac{z_{\frac{\alpha}{2}}\,\sigma}{E}\right]^2 = \left[\frac{2.58(2.6)}{0.5}\right]^2$

$$= 179.98 \text{ or } 180$$

19. $n = \left[\frac{z_{\frac{\alpha}{2}}\,\sigma}{E}\right]^2 = \left[\frac{1.65(900)}{300}\right]^2$

$$= 24.5 \text{ or } 25$$

20. $\hat{p} - (z_{\frac{\alpha}{2}})\sqrt{\frac{\hat{p}\hat{q}}{n}} < p < \hat{p} + (z_{\frac{\alpha}{2}})\sqrt{\frac{\hat{p}\hat{q}}{n}}$

$$\hat{p} = \frac{53}{75} = 0.707 \qquad \hat{q} = \frac{22}{75} = 0.293$$

$$0.71 - 1.96\sqrt{\frac{(0.707)(0.293)}{75}} < p <$$
$$0.71 + 1.96\sqrt{\frac{(0.707)(0.293)}{75}}$$

$$0.604 < p < 0.810$$

21. $\hat{p} - (z_{\frac{\alpha}{2}})\sqrt{\frac{\hat{p}\hat{q}}{n}} < p < \hat{p} + (z_{\frac{\alpha}{2}})\sqrt{\frac{\hat{p}\hat{q}}{n}}$

21. continued

$$0.36 - 1.65\sqrt{\frac{(0.36)(0.64)}{150}} < p < 0.36 + 1.65\sqrt{\frac{(0.36)(0.64)}{150}}$$

$$0.295 < p < 0.425$$

22. $\hat{p} - (z_{\frac{\alpha}{2}})\sqrt{\frac{\hat{p}\hat{q}}{n}} < p < \hat{p} + (z_{\frac{\alpha}{2}})\sqrt{\frac{\hat{p}\hat{q}}{n}}$

$$0.4444 - 1.96\sqrt{\frac{(0.4444)(0.5556)}{90}} < p < 0.4444 + 1.96\sqrt{\frac{(0.4444)(0.5556)}{90}}$$

$$0.342 < p < 0.547$$

23. $n = \hat{p}\,\hat{q}\left[\frac{z_{\frac{\alpha}{2}}}{E}\right]^2$

$$= (0.15)(0.85)\left[\frac{1.96}{0.03}\right]^2$$

$$= 544.22 \text{ or } 545$$

24. $\frac{(n-1)s^2}{\chi^2_{\text{right}}} < \sigma^2 < \frac{(n-1)s^2}{\chi^2_{\text{left}}}$

$$\frac{24(9)^2}{39.364} < \sigma^2 < \frac{24(9)^2}{12.401}$$

$$49.4 < \sigma^2 < 156.8$$

$$7 < \sigma < 13$$

25. $\frac{(n-1)s^2}{\chi^2_{\text{right}}} < \sigma^2 < \frac{(n-1)s^2}{\chi^2_{\text{left}}}$

$$\frac{26(6.8)^2}{38.885} < \sigma^2 < \frac{26(6.8)^2}{15.379}$$

$$30.9 < \sigma^2 < 78.2$$

$$5.6 < \sigma < 8.8$$

26. $\frac{(n-1)s^2}{\chi^2_{\text{right}}} < \sigma^2 < \frac{(n-1)s^2}{\chi^2_{\text{left}}}$

$$\frac{19(2.3)^2}{30.144} < \sigma^2 < \frac{19(2.3)^2}{10.177}$$

$$3.33 < \sigma^2 < 10$$

$$1.8 < \sigma < 3.2$$

Note: Graphs are not to scale and are intended to convey a general idea.

Answers may vary due to rounding.

EXERCISE SET 8-2

1.
The null hypothesis is a statistical hypothesis that states there is no difference between a parameter and a specific value or there is no difference between two parameters. The alternative hypothesis specifies a specific difference between a parameter and a specific value, or that there is a difference between two parameters. Examples will vary.

3.
A statistical test uses the data obtained from a sample to make a decision as to whether or not the null hypothesis should be rejected.

5.
The critical region is the region of values of the test-statistic that indicates a significant difference and the null hypothesis should be rejected. The non-critical region is the region of values of the test-statistic that indicates the difference was probably due to chance, and the null hypothesis should not be rejected.

7.
Type I is represented by α, type II is represented by β.

9.
A one-tailed test should be used when a specific direction, such as greater than or less than, is being hypothesized, whereas when no direction is specified, a two-tailed test should be used.

11.
Hypotheses can only be proved true when the entire population is used to compute the test statistic. In most cases, this is impossible.

12.
a. $+2.58$, -2.58

12a. continued

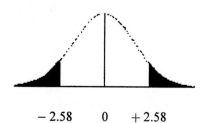

$-2.58 \qquad 0 \qquad +2.58$

b. $+1.65$

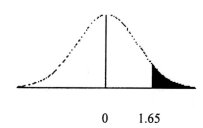

$0 \qquad 1.65$

c. -2.58

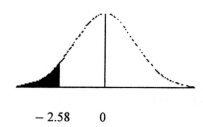

$-2.58 \qquad 0$

d. -1.28

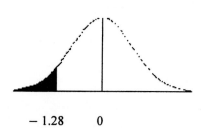

$-1.28 \qquad 0$

e. $+1.96$, -1.96

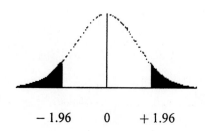

$-1.96 \qquad 0 \qquad +1.96$

12. continued

f. $+1.75$

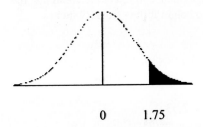

0   1.75

g. -2.33

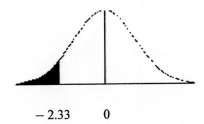

-2.33 0

h. $+1.65, -1.65$

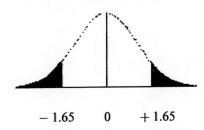

-1.65 0 $+1.65$

i. $+2.05$

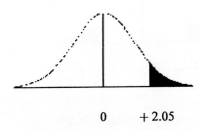

0    $+2.05$

j. $+2.33, -2.33$

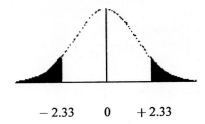

-2.33 0 $+2.33$

13.

a. H_0: $\mu = 36.3$ (claim)
      H_1: $\mu \neq 36.3$

b. H_0: $\mu = \$36,250$ (claim)
      H_1: $\mu \neq \$36,250$

c. H_0: $\mu \leq 27.6$ years
      H_1: $\mu > 27.6$ years   (claim)

d. H_0: $\mu \geq 72$
      H_1: $\mu < 72$ (claim)

e. H_0: $\mu \geq 100$
      H_1: $\mu < 100$ (claim)

f. H_0: $\mu = \$297.75$ (claim)
      H_1: $\mu \neq \$297.75$

g. H_0: $\mu \leq \$52.98$
      H_1: $\mu > \$52.98$ (claim)

h. H_0: $\mu \leq 300$ (claim)
      H_1: $\mu > 300$

i. H_0: $\mu \geq 3.6$ (claim)
      H_1: $\mu < 3.6$

EXERCISE SET 8-3

1.
H_0: $\mu = \$69.21$ (claim)
H_1: $\mu \neq \$69.21$

C. V. $= \pm 1.96$
$$z = \frac{\overline{X} - \mu}{\frac{\sigma}{\sqrt{n}}} = \frac{\$68.43 - \$69.21}{\frac{3.72}{\sqrt{30}}} = -1.15$$

-1.96 ↑   0 $+1.96$
    -1.15

Do not reject the null hypothesis. There is not enough evidence to reject the claim that the average cost of a hotel stay in Atlanta is $69.21.

3.
H_0: $\mu \leq \$24$ billion
$H1$: $\mu > \$24$ billion   (claim)

3. continued

C. V. $= +1.65$ $\overline{X} = \$31.5$ $s = \$28.7$

$z = \frac{\overline{X}-\mu}{\frac{\sigma}{\sqrt{n}}} = \frac{31.5-24}{\frac{28.7}{\sqrt{50}}} = 1.85$

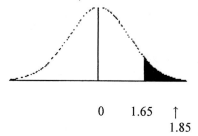

0   1.65   ↑
1.85

Reject the null hypothesis. There is enough evidence to support the claim that the average revenue exceeds $24 billion.

5.
H_0: $\mu \geq 14$
H_1: $\mu < 14$ (claim)

C. V. $= -2.33$

$z = \frac{\overline{X}-\mu}{\frac{s}{\sqrt{n}}} = \frac{11.8-14}{\frac{2.7}{\sqrt{36}}} = -4.89$

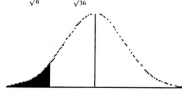

↑ -2.33 0
-4.89

Reject the null hypothesis. There is enough evidence to support the claim that the average age of the planes in the executive's airline is less than the national average.

7.
H_0: $\mu = 29$
H_1: $\mu \neq 29$ (claim)

C. V. $= \pm 1.96$ $\overline{X} = 29.45$ $s = 2.61$

$z = \frac{\overline{X}-\mu}{\frac{\sigma}{\sqrt{n}}} = \frac{29.45-29}{\frac{2.61}{\sqrt{30}}} = 0.944$

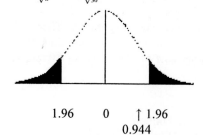

1.96   0   ↑ 1.96
0.944

7. continued

Do not reject the null hypothesis. There is enough evidence to reject the claim that the average height differs from 29 inches.

9.
H_0: $\mu \leq \$19,410$
H_1: $\mu > \$19,410$ (claim)

C. V. $= 2.33$

$z = \frac{\overline{X}-\mu}{\frac{\sigma}{\sqrt{n}}} = \frac{\$22,098-\$19,410}{\frac{6050}{\sqrt{40}}} = 2.81$

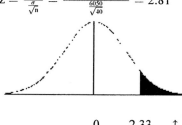

0   2.33   ↑
2.81

Reject the null hypothesis. There is enough evidence to support the claim that the average tuition cost has increased.

11.
H_0: $\mu = 125$
H_1: $\mu \neq 125$ (claim)

C. V. $= \pm 2.58$

$z = \frac{\overline{X}-\mu}{\frac{\sigma}{\sqrt{n}}} = \frac{110-125}{\frac{30}{\sqrt{35}}} = -2.96$

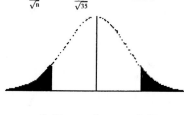

-2.58 0   2.58
↑ -2.96

Reject the null hypothesis. There is a significant difference in the average number of guests.

13.
H_0: $\mu = \$24.44$
H_1: $\mu \neq \$24.44$ (claim)

C. V. $= \pm 2.33$

$z = \frac{\overline{X}-\mu}{\frac{s}{\sqrt{n}}} = \frac{22.97-24.44}{\frac{3.70}{\sqrt{33}}} = -2.28$

13. continued

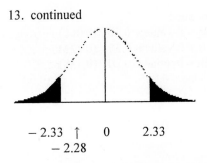

$-2.33 \uparrow \quad 0 \qquad 2.33$
$\quad -2.28$

Do not reject the null hypothesis. There is not enough evidence to support the claim that the amount spent at a local mall is not equal to the national average of $24.44.

15.
a. Do not reject.
b. Do not reject.
c. Do not reject.
d. Reject
e. Reject

17.
H_0: $\mu \geq 264$
H_1: $\mu < 264$ (claim)
$z = \frac{\overline{X}-\mu}{\frac{\sigma}{\sqrt{n}}} = \frac{262.3-264}{\frac{3}{\sqrt{20}}} = -2.53$

The area corresponding to $z = 2.53$ is 0.4943. The P-value is $0.5 - 0.4943 = 0.0057$. The decision is to reject the null hypothesis since $0.0057 < 0.01$. There is enough evidence to support the claim that the average stopping distance is less than 264 feet.

19.
H_0: $\mu \leq 84$
H_1: $\mu > 84$ (claim)
$z = \frac{\overline{X}-\mu}{\frac{\sigma}{\sqrt{n}}} = \frac{85.1-84}{\frac{10}{\sqrt{100}}} = 1.1$

The area corresponding to $z = 1.1$ is 0.3643. The P-value is $0.5 - 0.3643 = 0.1357$. The decision is do not reject the null hypothesis since $0.1357 > 0.01$. There is not enough evidence to support the claim that the average lifetime of the television sets is greater than 84 months.

21.
H_0: $\mu = 6.32$ (claim)
H_1: $\mu \neq 6.32$
$z = \frac{\overline{X}-\mu}{\frac{\sigma}{\sqrt{n}}} = \frac{6.51-6.32}{\frac{0.54}{\sqrt{50}}} = 2.49$

21. continued
The area corresponding to $z = 2.49$ is 0.4936. To get the P-value, subtract 0.4936 from 0.5 and then multiply by 2 since this is a two-tailed test.
$2(0.5 - 0.4936) = 2(0.0064) = 0.0128$
The decision is to reject the null hypothesis since $0.0128 < 0.05$. There is enough evidence to reject the claim that the average wage is $6.32.

23.
H_0: $\mu = 30,000$ (claim)
H_1: $\mu \neq 30,000$
$z = \frac{\overline{X}-\mu}{\frac{s}{\sqrt{n}}} = \frac{30,456-30,000}{\frac{1684}{\sqrt{40}}} = 1.71$

The area corresponding to $z = 1.71$ is 0.4564. The P-value is $2(0.5 - 0.4564) = 2(0.0436) = 0.0872$.

The decision is to reject the null hypothesis at $\alpha = 0.10$ since $0.0872 < 0.10$. The conclusion is that there is enough evidence to reject the claim that customers are adhering to the recommendation. A 0.10 significance level is probably appropriate since there is little consequence of a Type I error. The dealer would be advised to increase efforts to make its customers aware of the service recommendation.

25.
H_0: $\mu \geq 10$
H_1: $\mu < 10$ (claim)

$\overline{X} = 5.025 \quad s = 3.63$
$z = \frac{\overline{X}-\mu}{\frac{s}{\sqrt{n}}} = \frac{5.025-10}{\frac{3.63}{\sqrt{40}}} = -8.67$

The area corresponding to 8.67 is greater than 0.4999. The P-value is $0.5 - 0.4999 < 0.0001$. Since $0.0001 < 0.05$, the decision is to reject the null hypothesis. There is enough evidence to support the claim that the average number of days missed per year is less than 10.

27.
The mean and standard deviation are found as follows:

27. continued

	f	X_m	$f \cdot X_m$	$f \cdot X_m^2$
8.35 - 8.43	2	8.39	16.78	140.7842
8.44 - 8.52	6	8.48	50.88	431.4624
8.53 - 8.61	12	8.57	102.84	881.3388
8.62 - 8.70	18	8.66	155.88	1349.9208
8.71 - 8.79	10	8.75	87.5	765.625
8.80 - 8.88	2	8.84	17.68	156.2912
	50		431.56	3725.4224

$$\overline{X} = \frac{\sum f \cdot X_m}{n} = \frac{431.56}{50} = 8.63$$

$$s = \sqrt{\frac{\sum f \cdot X_m^2 - \frac{(\sum f \cdot X_m)^2}{n}}{n-1}} = \sqrt{\frac{3725.4224 - \frac{(431.56)^2}{50}}{49}}$$

$$= 0.105$$

H_0: $\mu = 8.65$ (claim)
H_1: $\mu \neq 8.65$

C. V. $= \pm 1.96$
$$z = \frac{\overline{X} - \mu}{\frac{s}{\sqrt{n}}} = \frac{8.63 - 8.65}{\frac{0.105}{\sqrt{50}}} = -1.35$$

Do not reject the null hypothesis. There is not enough evidence to reject the claim that the average hourly wage of the employees is $8.65.

EXERCISE SET 8-4

1.
It is bell-shaped, symmetric about the mean, and it never touches the x axis. The mean, median, and mode are all equal to 0 and they are located at the center of the distribution. The t distribution differs from the standard normal distribution in that it is a family of curves, the variance is greater than one, and as the degrees of freedom increase the t distribution approaches the standard normal distribution.

3.
a. d. f. = 9   C. V. = + 1.833
b. d. f. = 17   C. V. = ± 1.740
c. d. f. = 5   C. V. = − 3.365
d. d. f. = 8   C. V. = + 2.306
e. d. f. = 14   C. V. = ± 2.145
f. d. f. = 22   C. V. = − 2.819
g. d. f. = 27   C. V. = ± 2.771
h. d. f. = 16   C. V. = ± 2.583

4.
a. 0.01 < P-value < 0.025 (0.018)

4. continued
b. 0.05 < P-value < 0.10 (0.062)
c. 0.10 < P-value < 0.25 (0.123)
d. 0.10 < P-value < 0.20 (0.138)
e. P-value < 0.005   (0.003)
f. 0.10 < P-value < 0.25 (0.158)
g. P-value = 0.05   (0.05)
h. P-value > 0.25   (0.261)

5.
H_0: $\mu \geq 11.52$
H_1: $\mu < 11.52$ (claim)

C. V. $= -1.833$ d. f. $= 6$

$$t = \frac{\overline{X} - \mu}{\frac{s}{\sqrt{n}}} = \frac{7.42 - 11.52}{\frac{1.3}{\sqrt{10}}} = -9.97$$

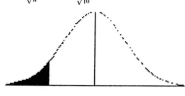

↑ − 1.833   0
− 9.97

Reject the null hypothesis. There is enough evidence to support the claim that the rainfall is below average.

7.
H_0: $\mu = \$40,000$
H_1: $\mu \neq \$40,000$ (claim)

C. V. $= \pm 2.093$ d. f. $= 19$
$$t = \frac{\overline{X} - \mu}{\frac{s}{\sqrt{n}}} = \frac{43,228 - 40,000}{\frac{4000}{\sqrt{20}}} = 3.61$$

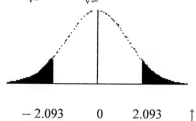

− 2.093   0   2.093   ↑
3.61

Reject the null hypothesis. There is enough evidence to support the claim that the average salary is not $40,000.

9.
H_0: $\mu \geq 700$ (claim)
H_1: $\mu < 700$
$\overline{X} = 606.5$ s = 109.1

9. continued

C. V. $= -2.262$ d. f. $= 9$

$t = \frac{\overline{X} - \mu}{\frac{s}{\sqrt{n}}} = \frac{606.5 - 700}{\frac{109.1}{\sqrt{10}}} = -2.71$

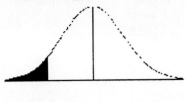

↑ -2.262 0
-2.71

Reject the null hypothesis. There is enough evidence to reject the claim that the average height of the buildings is at least 700 feet.

11.

H_0: $\mu \leq \$13,252$
H_1: $\mu > \$13,252$ (claim)

C. V. $= 2.539$ d. f. $= 19$

$t = \frac{\overline{X} - \mu}{\frac{s}{\sqrt{n}}} = \frac{\$15,560 - \$13,252}{\frac{\$3500}{\sqrt{19}}} = 2.95$

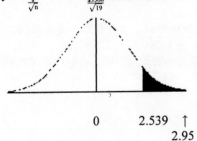

0   2.539 ↑
                2.95

Reject the null hypothesis. There is enough evidence to support the claim that the average tuition cost has increased.

13.

H_0: $\mu \leq \$54.8$
H_1: $\mu > \$54.8$ (claim)

C. V. $= 1.761$ d. f. $= 14$

$t = \frac{\overline{X} - \mu}{\frac{s}{\sqrt{n}}} = \frac{\$62.3 - \$54.8}{\frac{\$9.5}{\sqrt{15}}} = 3.06$

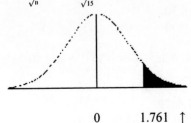

0   1.761 ↑
                3.06

13. continued

Reject the null hypothesis. There is enough evidence to support the claim that the cost to produce an action movie is more than $54.8.

15.

H_0: $\mu = 132$ min.   (claim)
H_1: $\mu \neq 132$ min.

C. V. $= \pm 2.365$ d. f. $= 7$

$t = \frac{\overline{X} - \mu}{\frac{s}{\sqrt{n}}} = \frac{125 - 132}{\frac{11}{\sqrt{8}}} = -1.80$

Do not reject the null hypothesis. There is enough evidence to support the claim that the average show time is 132 minutes, or 2 hours and 12 minutes.

17.

H_0: $\mu = 75$ (claim)
H_1: $\mu \neq 75$
$\overline{X} = 70.85$ $s = 6.56$
d. f. $= 19$
$0.01 < $ P-value $ < 0.02$ (0.011)
$t = \frac{\overline{X} - \mu}{\frac{s}{\sqrt{n}}} = \frac{70.85 - 75}{\frac{6.56}{\sqrt{20}}} = -2.83$

Since P-value > 0.01, do not reject the null hypothesis. There is not enough evidence to reject the claim that the average score on the real estate exam is 75.

19.

H_0: $\mu = \$15,000$
H_1: $\mu \neq \$15,000$ (claim)

$\overline{X} = \$14,347.17$ $s = \$2048.54$
d. f. $= 11$ C. V. $= \pm 2.201$

$t = \frac{\overline{X} - \mu}{\frac{s}{\sqrt{n}}} = \frac{\$14,347.17 - \$15,000}{\frac{\$2048.54}{\sqrt{12}}} = -1.10$

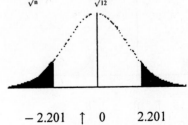

-2.201 ↑ 0   2.201
          -1.10

Do not reject the null hypothesis. There is not enough evidence to say that the average stipend differs from $15,000.

EXERCISE SET 8-5

1.
Answers will vary.

3.
$np \geq 5$ and $nq \geq 5$

5.
H_0: $p = 0.647$
H_1: $p \neq 0.647$ (claim)

$\hat{p} = \frac{92}{150} = 0.613$ $p = 0.647$ $q = 0.353$
C. V. $= \pm 2.58$

$z = \frac{\hat{p} - p}{\sqrt{\frac{pq}{n}}} = \frac{0.613 - 0.647}{\sqrt{\frac{(0.647)(0.353)}{150}}} = -0.86$

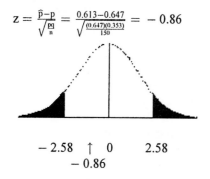

-2.58 ↑ 0   2.58
      -0.86

Do not reject the null hypothesis. There is not enough evidence to support the claim that the proportion of homeowners is different from 64.7%.

7.
H_0: $p = 0.40$
H_1: $p \neq 0.40$ (claim)

$\hat{p} = \frac{65}{180} = 0.361$ $p = 0.40$ $q = 0.60$
C. V. $= \pm 2.58$
$z = \frac{\hat{p} - p}{\sqrt{\frac{pq}{n}}} = \frac{0.361 - 0.40}{\sqrt{\frac{(0.40)(0.60)}{180}}} = -1.07$

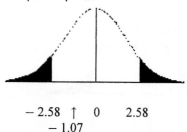

-2.58 ↑ 0   2.58
      -1.07

Do not reject the null hypothesis. There is not enough evidence to conclude that the proportion differs from 40%.

9.
H_0: $p = 0.63$ (claim)

9. continued
H_1: $p \neq 0.63$

$\hat{p} = \frac{85}{143} = 0.5944$ $p = 0.63$ $q = 0.37$
C. V. $= \pm 1.96$
$z = \frac{\hat{p} - p}{\sqrt{\frac{pq}{n}}} = \frac{0.5944 - 0.63}{\sqrt{\frac{(0.63)(0.37)}{143}}} = -0.88$

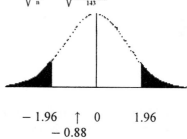

-1.96 ↑ 0   1.96
      -0.88

Do not reject the null hypothesis. There is not enough evidence to reject the claim that the percentage is the same.

11.
H_0: $p = 0.54$
H_1: $p \neq 0.54$ (claim)

$\hat{p} = \frac{14}{30} = 0.4667$ $p = 0.54$ $q = 0.46$
C. V. $= \pm 1.96$
$z = \frac{\hat{p} - p}{\sqrt{\frac{pq}{n}}} = \frac{0.4667 - 0.54}{\sqrt{\frac{(0.54)(0.46)}{30}}} = -0.81$

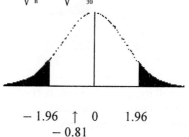

-1.96 ↑ 0   1.96
      -0.81

Do not reject the null hypothesis. There is not enough evidence to reject the claim that 54% of fatal car/truck accidents are caused by driver error.

13.
H_0: $p \leq 0.30$
H_1: $p > 0.30$ (claim)

$\hat{p} = \frac{72}{200} = 0.36$ $p = 0.30$ $q = 0.70$
$z = \frac{\hat{p} - p}{\sqrt{\frac{pq}{n}}} = \frac{0.36 - 0.30}{\sqrt{\frac{(0.30)(0.70)}{200}}} = 1.85$
Area $= 0.4678$
P-value $= 0.5 - 0.4678 = 0.0322$
Since P-value < 0.05, reject the null hypothesis. There is enough evidence to

13. continued
support the claim that more than 30% of the customers have at least two telephones.

15.
H_0: $p = 0.18$ (claim)
H_1: $p \neq 0.18$

$\hat{p} = \frac{50}{300} = 0.1667$ $p = 0.18$ $q = 0.82$
$z = \frac{\hat{p}-p}{\sqrt{\frac{pq}{n}}} = \frac{0.1667-0.18}{\sqrt{\frac{(0.18)(0.82)}{300}}} = -0.60$
Area $= 0.2257$
P-value $= 2(0.5 - 0.2257) = 0.5486$
Since P-value > 0.01, do not reject the null hypothesis. There is not enough evidence to reject the claim that 18% of all high school students smoke at least a pack of cigarettes a day.

17.
H_0: $p = 0.67$
H_1: $p \neq 0.67$ (claim)

$\hat{p} = \frac{82}{100} = 0.82$ $p = 0.67$ $q = 0.33$
C. V. $= \pm 1.96$
$z = \frac{\hat{p}-p}{\sqrt{\frac{pq}{n}}} = \frac{0.82-0.67}{\sqrt{\frac{(0.67)(0.33)}{100}}} = 3.19$

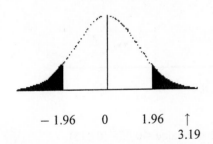

-1.96 0 1.96 $\uparrow$
3.19

Reject the null hypothesis. There is enough evidence to support the claim that the percentage is not 67%.

19.
H_0: $p \geq 0.576$
H_1: $p < 0.576$ (claim)

$\hat{p} = \frac{17}{36} = 0.472$ $p = 0.576$ $q = 0.424$
C. V. $= -1.65$
$z = \frac{\hat{p}-p}{\sqrt{\frac{pq}{n}}} = \frac{0.472-0.576}{\sqrt{\frac{(0.576)(0.424)}{36}}} = -1.26$

19. continued

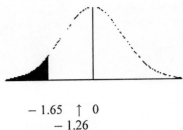

-1.65 $\uparrow$ 0
-1.26

Do not reject the null hypothesis. There is not enough evidence to support the claim that the percentage of injuries during practice is below 57.6%.

21.
$P(X = 3, p = 0.5, n = 9) = 0.164$
Since $0.164 > \frac{0.10}{2}$ or 0.05, the conclusion that the coin is not balanced is probably false. Note that α must be split since this is a two-tailed test.

23.
$z = \frac{X-\mu}{\sigma}$

$z = \frac{X-np}{\sqrt{npq}}$

$z = \frac{\frac{X}{n} - \frac{np}{n}}{\frac{1}{n}\sqrt{npq}}$

$z = \frac{\frac{X}{n} - \frac{np}{n}}{\sqrt{\frac{npq}{n^2}}}$

$z = \frac{\hat{p}-p}{\sqrt{\frac{pq}{n}}}$

EXERCISE SET 8-6

1.
a. H_0: $\sigma^2 \leq 225$
   H_1: $\sigma^2 > 225$

C. V. $= 27.587$ d. f. $= 17$

0 27.587

b. H_0: $\sigma^2 \geq 225$
   H_1: $\sigma^2 < 225$

1b. continued
C. V. = 14.042   d. f. = 22

0   14.042

c. H_0: $\sigma^2 = 225$
   H_1: $\sigma^2 \neq 225$

C. V. = 5.629, 26.119   d. f. = 14

0   5.629   26.119

d. H_0: $\sigma^2 = 225$
   H_1: $\sigma^2 \neq 225$

C. V. = 2.167, 14.067   d. f. = 7

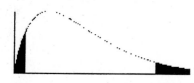

0   2.167   14.067

e. H_0: $\sigma^2 \leq 225$
   H_1: $\sigma^2 > 225$

C. V. = 32.000   d. f. = 16

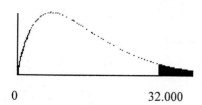

0   32.000

f. H_0: $\sigma^2 \geq 225$
   H_1: $\sigma^2 < 225$

C. V. = 8.907   d. f. = 19

1f. continued

0   8.907

g. H_0: $\sigma^2 = 225$
   H_1: $\sigma^2 \neq 225$

C. V. = 3.074, 28.299   d. f. = 12

0   3.074   28.299

h. H_0: $\sigma^2 \geq 225$
   H_1: $\sigma^2 < 225$

C. V. = 15.308   d. f. = 28

0   15.308

2.
a. 0.01 < P-value < 0.025 (0.015)
b. 0.005 < P-value < 0.01 (0.006)
c. 0.01 < P-value < 0.025 (0.012)
d. P-value < 0.005   (0.003)
e. 0.025 < P-value < 0.05 (0.037)
f. 0.05 < P-value < 0.10   (0.088)
g. 0.05 < P-value < 0.10   (0.066)
h. P-value < 0.01   (0.007)

3.
H_0: $\sigma = 60$ (claim)
H_1: $\sigma \neq 60$

C. V. = 8.672, 27.587 $\alpha = 0.10$
d. f. = 17
s = 64.6
$\chi^2 = \frac{(n-1)s^2}{\sigma^2} = \frac{(18-1)(64.6)^2}{(60)^2} = 19.707$

3. continued

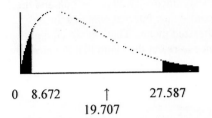

0   8.672   ↑   27.587
                    19.707

Do not reject the null hypothesis. There is not enough evidence to reject the claim that the standard deviation is 60.

5.
H_0: $\sigma^2 \le 25$ (claim)
H_1: $\sigma^2 > 25$

C. V. = 27.204 $\alpha = 0.10$ d. f. = 19

$\chi^2 = \frac{(n-1)s^2}{\sigma^2} = \frac{(20-1)(36)}{25} = 27.36$

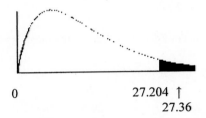

0   27.204 ↑
                              27.36

Reject the null hypothesis. There is enough evidence to reject the claim that the variance is less than or equal to 25.

7.
H_0: $\sigma \le 1.2$ (claim)
H_1: $\sigma > 1.2$

$\alpha = 0.01$ d. f. = 14
$\chi^2 = \frac{(n-1)s^2}{\sigma^2} = \frac{(15-1)(1.8)^2}{(1.2)^2} = 31.5$
P-value < 0.005 (0.0047)

Since P-value < 0.01, reject the null hypothesis. There is enough evidence to reject the claim that the standard deviation is less than or equal to 1.2 minutes.

9.
H_0: $\sigma \le 20$
H_1: $\sigma > 20$ (claim)

s = 35.11
C. V. = 36.191 $\alpha = 0.01$ d. f. = 19

9. continued
$\chi^2 = \frac{(n-1)s^2}{\sigma^2} = \frac{(20-1)(35.11)^2}{20^2} = 58.55$

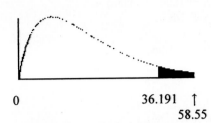

0   36.191 ↑
                                58.55

Reject the null hypothesis. There is enough evidence to support the claim that the standard deviation is more than 20 calories.

11.
H_0: $\sigma \le 100$
H_1: $\sigma > 100$ (claim)

C. V. = 124.342 $\alpha = 0.05$ d. f. = 299
$\chi^2 = \frac{(n-1)s^2}{\sigma^2} = \frac{(300-1)(110)^2}{100^2} = 361.79$

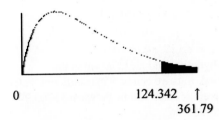

0   124.342   ↑
                                  361.79

Reject the null hypothesis. There is enough evidence to support the claim that the standard deviation is more than 100.

13.
H_0: $\sigma \le 25$
H_1: $\sigma > 25$ (claim)

C. V. = 22.362 $\alpha = 0.05$ d. f. = 13
$\chi^2 = \frac{(n-1)s^2}{\sigma^2} = \frac{(14-1)(6.74)^2}{25} = 23.622$

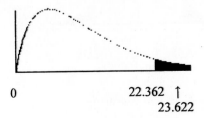

0   22.362 ↑
                                23.622

Reject the null hypothesis. There is enough evidence to support the claim that the variance is greater than 25.

EXERCISE SET 8-7

1.
H_0: $\mu = 1800$ (claim)
H_1: $\mu \neq 1800$

C. V. $= \pm 1.96$
$z = \frac{\overline{X}-\mu}{\frac{\sigma}{\sqrt{n}}} = \frac{1830-1800}{\frac{200}{\sqrt{10}}} = 0.47$

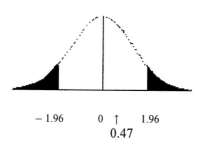

$$-1.96 \qquad 0 \uparrow \quad 1.96$$
$$0.47$$

The 95% confidence interval of the mean is:
$$\overline{X} - z_{\frac{\alpha}{2}} \frac{\sigma}{\sqrt{n}} < \mu < \overline{X} + z_{\frac{\alpha}{2}} \frac{\sigma}{\sqrt{n}}$$

$$1830 - 1.96\left(\frac{200}{\sqrt{10}}\right) < \mu <$$
$$1830 + 1.96\left(\frac{200}{\sqrt{10}}\right)$$
$$1706.04 < \mu < 1953.96$$

The hypothesized mean is within the interval, thus we can be 95% confident that the average sales will be between $1706.94 and $1953.96.

3.
H_0: $\mu = 86$ (claim)
H_1: $\mu \neq 86$

C. V. $= \pm 2.58$
$z = \frac{\overline{X}-\mu}{\frac{\sigma}{\sqrt{n}}} = \frac{84-86}{\frac{6}{\sqrt{15}}} = -1.29$

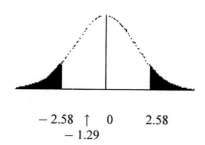

$$-2.58 \uparrow \quad 0 \qquad 2.58$$
$$-1.29$$

$$\overline{X} - z_{\frac{\alpha}{2}} \frac{\sigma}{\sqrt{n}} < \mu < \overline{X} + z_{\frac{\alpha}{2}} \frac{\sigma}{\sqrt{n}}$$

$$84 - 2.58 \cdot \frac{6}{\sqrt{15}} < \mu < 84 + 1.58 \cdot \frac{6}{\sqrt{15}}$$

$$80.00 < \mu < 88.00$$

3. continued
The decision is do not reject the null hypothesis since $-1.29 > -2.58$ and the 99% confidence interval contains the hypothesized mean. There is not enough evidence to reject the claim that the monthly maintenance is $86.

5.
H_0: $\mu = 22$
H_1: $\mu \neq 22$ (claim)

C. V. $= \pm 2.58$
$z = \frac{\overline{X}-\mu}{\frac{\sigma}{\sqrt{n}}} = \frac{20.8-22}{\frac{4}{\sqrt{60}}} = -2.32$

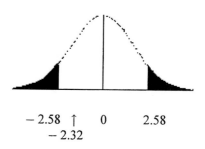

$$-2.58 \uparrow \quad 0 \qquad 2.58$$
$$-2.32$$

The 99% confidence interval of the mean is:

$$\overline{X} - z_{\frac{\alpha}{2}} \frac{\sigma}{\sqrt{n}} < \mu < \overline{X} + z_{\frac{\alpha}{2}} \frac{\sigma}{\sqrt{n}}$$

$$20.8 - 2.58 \cdot \frac{4}{\sqrt{60}} < \mu < 20.8 + 2.58 \cdot \frac{4}{\sqrt{60}}$$

$$19.47 < \mu < 22.13$$

The decision is do not reject the null hypothesis since $-2.32 > -2.58$ and the 99% confidence interval does contain the hypothesized mean of 22. The conclusion is that there is not enough evidence to support the claim that the average studying time has changed.

7.
The power of a statistical test is the probability of rejecting the null hypothesis when it is false.

9.
The power of a test can be increased by increasing α or selecting a larger sample size.

REVIEW EXERCISES - CHAPTER 8

1.
H_0: $\mu = 98°$ (claim)
H_1: $\mu \neq 98°$

C. V. $= \pm 1.96$
$z = \frac{\overline{X}-\mu}{\frac{s}{\sqrt{n}}} = \frac{95.8-98}{\frac{7.71}{\sqrt{50}}} = -2.02$

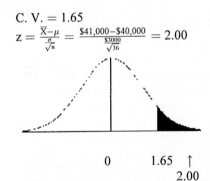

↑ -1.96 0   1.96
-2.02

Reject the null hypothesis. There is enough evidence to reject the claim that the average high temperature is 98°.

3.
H_0: $\mu \leq \$40,000$
H_1: $\mu > \$40,000$ (claim)

C. V. $= 1.65$
$z = \frac{\overline{X}-\mu}{\frac{\sigma}{\sqrt{n}}} = \frac{\$41,000-\$40,000}{\frac{\$3000}{\sqrt{36}}} = 2.00$

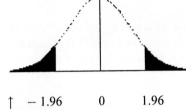

0   1.65   ↑
           2.00

Reject the null hypothesis. There is enough evidence to support the claim that the average salary is more than $40,000.

5.
H_0: $\mu \leq 67$
H_1: $\mu > 67$ (claim)

C. V. $= 1.383$ d. f. $= 9$
$t = \frac{\overline{X}-\mu}{\frac{s}{\sqrt{n}}} = \frac{69.6-67.0}{\frac{1.1}{\sqrt{10}}} = 7.47$

5. continued

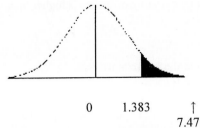

0   1.383   ↑
                  7.47

Reject the null hypothesis. There is enough evidence to support the claim that 1995 was warmer than average.

7.
H_0: $\mu = 6$ (claim)
H_1: $\mu \neq 6$

C. V. $= \pm 2.821$ $\overline{X} = 8.42$ $s = 4.17$
$t = \frac{\overline{X}-\mu}{\frac{s}{\sqrt{n}}} = \frac{8.42-6}{\frac{4.17}{\sqrt{10}}} = 1.835$

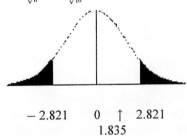

-2.821 0   ↑   2.821
               1.835

Do not reject the null hypothesis. There is not enough evidence to support the claim that the average attendance has changed.

9.
H_0: $p \leq 0.602$
H_1: $p > 0.602$ (claim)

C. V. $= 1.65$

$\hat{p} = 0.65$ $p = 0.602$ $q = 0.398$
$z = \frac{\hat{p}-p}{\sqrt{\frac{pq}{n}}} = \frac{0.65-0.602}{\sqrt{\frac{(0.602)(0.398)}{400}}} = 1.96$

0   1.65   ↑
              1.96

Reject the null hypothesis. There is enough evidence to support the claim that the

9. continued

percentage of drug offenders is higher than 60.2%.

11.

H_0: $p = 0.65$ (claim)

H_1: $p \neq 0.65$

$\hat{p} = \frac{57}{80} = 0.7125$ $p = 0.65$ $q = 0.35$

$z = \frac{\hat{p} - p}{\sqrt{\frac{pq}{n}}} = \frac{0.7125 - 0.65}{\sqrt{\frac{(0.65)(0.35)}{80}}} = 1.17$

Area = 0.3790

P-value = $2(0.5 - 0.3790) = 0.242$

Since P-value > 0.05, do not reject the null hypothesis. There is not enough evidence to reject the claim that 65% of the teenagers own their own radios.

13.

H_0: $\mu \geq 10$

H_1: $\mu < 10$ (claim)

$z = \frac{\overline{X} - \mu}{\frac{\sigma}{\sqrt{n}}} = \frac{9.25 - 10}{\frac{2}{\sqrt{35}}} = -2.22$

Area = 0.4868

P-value = $0.5 - 0.4699 = 0.0132$

Since $0.0132 < 0.05$, reject the null hypothesis. The conclusion is that there is enough evidence to support the claim that the average time is less than 10 minutes.

15.

H_0: $\sigma \geq 4.3$ (claim)

H_1: $\sigma < 4.3$

d. f. = 19

$\chi^2 = \frac{(n-1)s^2}{\sigma^2} = \frac{(20-1)(2.6)^2}{(4.3^2} = 6.95$

$0.005 < $ P-value $ < 0.01$ (0.006)

Since P-value < 0.05, reject the null hypothesis. There is enough evidence to reject the claim that the standard deviation is greater than or equal to 4.3 miles per gallon.

17.

H_0: $\sigma = 18$ (claim)

H_1: $\sigma \neq 18$

C. V. = 11.143 and 0.484   d. f. = 4

$\chi^2 = \frac{(n-1)s^2}{\sigma^2} = \frac{(5-1)(21)^2}{18^2} = 5.44$

17. continued

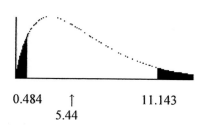

0.484   ↑   11.143

5.44

Do not reject the null hypothesis. There is not enough evidence to reject the claim that the standard deviation is 21 minutes.

19.

H_0: $\mu = 4$

H_1: $\mu \neq 4$ (claim)

C. V. = ± 2.58

$z = \frac{\overline{X} - \mu}{\frac{s}{\sqrt{n}}} = \frac{4.2 - 4}{\frac{0.6}{\sqrt{20}}} = 1.49$

The 99% confidence interval of the mean is:

$$\overline{X} - z_{\frac{\alpha}{2}} \frac{\sigma}{\sqrt{n}} < \mu < \overline{X} + z_{\frac{\alpha}{2}} \frac{\sigma}{\sqrt{n}}$$

$4.2 - 2.58 \cdot \frac{0.6}{\sqrt{20}} < \mu < 4.2 + 2.58 \cdot \frac{0.6}{\sqrt{20}}$

$3.85 < \mu < 4.55$

The decision is do not reject the null hypothesis since $1.49 < 2.58$ and the confidence interval does contain the hypothesized mean of 4. There is not enough evidence to support the claim that the growth has changed.

CHAPTER 8 QUIZ

1.   True

2.   True

3.   False, the critical value separates the critical region from the noncritical region.

4.   True

5.   False, it can be one-tailed or two-tailed.

6.   b.

7.   d.

8.   c.

9.   b.

10. type I

11. β

12. statistical hypothesis

13. right

14. n $-$ 1

15. H_0: $\mu = 28.6$ (claim)
H_1: $\mu \neq 28.6$
C. V. $= \pm 1.96$
$z = 2.14$
Reject the null hypothesis. There is enough evidence to reject the claim that the average age is 28.6.

16. H_0: $\mu = \$6,500$ (claim)
H_1: $\mu \neq \$6,500$
C. V. $= \pm 1.96$
$z = 5.27$
Reject the null hypothesis. There is enough evidence to reject the agent's claim.

17. H_0: $\mu \leq 8$
H_1: $\mu > 8$ (claim)
C. V. $= 1.65$
$z = 6.00$
Reject the null hypothesis. There is enough evidence to support the claim that the average number of sticks is greater than 8.

18. H_0: $\mu = 21$ (claim)
H_1: $\mu \neq 21$
C. V. $= \pm 2.921$
$t = -2.06$
Do not reject the null hypothesis. There is not enough evidence to reject the claim that the average number of dropouts is 21.

19. H_0: $\mu \geq 67$
H_1: $\mu < 67$ (claim)
$t = -3.1568$
P-value < 0.005 (0.003)
Since P-value < 0.05, reject the null hypothesis. There is enough evidence to support the claim that the average height is less than 67 inches.

20. H_0: $\mu \geq 12.4$
H_1: $\mu < 12.4$ (claim)
C. V. $= -1.345$
$t = -0.328$
Reject the null hypothesis. There is enough evidence to support the claim that the average is less than what the company claimed.

21. H_0: $\mu \leq 63.5$
H_1: $\mu > 63.5$ (claim)
$t = 0.47075$
P-value > 0.25 (0.322)
Since P-value > 0.05, do not reject the null hypothesis. There is not enough evidence to

21. continued
support the claim that the average is greater than 63.5.

22. H_0: $\mu = 26$ (claim)
H_1: $\mu \neq 26$
C. V. $= \pm 2.492$
$t = -1.5$
Do not reject the null hypothesis. There is not enough evidence to reject the claim that the average age is 26.

23. H_0: $p \geq 0.25$ (claim)
H_1: $p < 0.25$
$\mu = 25$ $\sigma = 4.33$
C. V. $= -1.65$
$z = -0.6928$
Do not reject the null hypothesis. There is not enough evidence to reject the claim that the proportion is at least 0.25.

24. H_0: $p \geq 0.55$ (claim)
H_1: $p < 0.55$
$\mu = 44$ $\sigma = 4.45$
C. V. $= -1.28$
$z = -0.899$
Do not reject the null hypothesis. There is not enough evidence to reject the dietitian's claim.

25. H_0: $p = 0.7$ (claim)
H_1: $p \neq 0.7$
$\mu = 21$ $\sigma = 2.51$
C. V. $= \pm 2.33$
$z = 0.7968$
Do not reject the null hypothesis. There is not enough evidence to reject the claim that the proportion is 0.7.

26. H_0: $p = 0.75$ (claim)
H_1: $p \neq 0.75$
$\mu = 45$ $\sigma = 3.35$
C. V. $= \pm 2.58$
$z = 2.6833$
Reject the null hypothesis. there is enough evidence to reject the claim.

27. The area corresponding to $z = 2.14$ is 0.4838.
P-value $= 2(0.5 - 0.4838) = 0.0324$

28. The area corresponding to $z = 5.27$ is greater than 0.4999.
Thus, P-value $\leq 2(0.5 - 0.4999) \leq 0.0002$.
(Note: Calculators give 0.0001)

29. H_0: $\sigma \leq 6$
H_1: $\sigma > 6$ (claim)
C. V. = 36.415
$\chi^2 = 54$
Reject the null hypothesis. There is enough
evidence to support the claim that the
standard deviation is more than 6 pages.

30. H_0: $\sigma = 8$ (claim)
H_1: $\sigma \neq 8$
C. V. = 27.991, 79.490
$\chi^2 = 33.2$
Do not reject the null hypothesis. There is
not enough evidence to reject the claim that
$\sigma = 8$.

31. H_0: $\sigma \geq 2.3$
H_1: $\sigma < 2.3$ (claim)
C. V. = 10.117
$\chi^2 = 13$
Reject the null hypothesis. There is enough
evidence to support the claim that the
standard deviation is less than 2.3.

32. H_0: $\sigma = 9$ (claim)
H_1: $\sigma \neq 9$
$\chi^2 = 13.4$
P-value > 0.20 (0.290)
Since P-value > 0.05, do not reject the null
hypothesis. There is not enough evidence to
reject the claim that $\sigma = 9$.

33. $28.3 < \mu < 30.1$

34. $\$6562.81 < \mu < \$6,637.19$

Note: Graphs are not to scale and are intended to convey a general idea. Answers may vary due to rounding, TI-83's, or computer programs.

EXERCISE SET 9-2

1.
Testing a single mean involves comparing a sample mean to a specific value such as $\mu = 100$; whereas testing the difference between means means comparing the means of two samples such as $\mu_1 = \mu_2$.

3.
The populations must be independent of each other and they must be normally distributed. s_1 and s_2 can be used in place of σ_1 and σ_2 when σ_1 and σ_2 are unknown and both samples are each greater than or equal to 30.

5.
H_0: $\mu_1 = \mu_2$ (claim)
H_1: $\mu_1 \neq \mu_2$

C. V. $= \pm 2.58$

$\overline{X}_1 = 662.6111$ $\overline{X}_2 = 758.875$
$s_1 = 449.8703$ $s_2 = 474.1258$

$$z = \frac{(\overline{X}_1 - \overline{X}_2) - (\mu_1 - \mu_2)}{\sqrt{\frac{\sigma_1^2}{n_1} + \frac{\sigma_2^2}{n_2}}} = \frac{(662.6111 - 758.875) - 0}{\sqrt{\frac{449.8703^2}{36} + \frac{474.1258^2}{36}}} =$$
$z = -0.88$

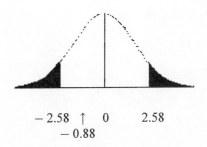

$-2.58 \uparrow\ 0 \quad 2.58$
$\quad -0.88$

Do not reject the null hypothesis. There is not enough evidence to reject the claim that the average lengths of the rivers is the same.

7.
H_0: $\mu_1 \leq \mu_2$
H_1: $\mu_1 > \mu_2$ (claim)

C. V. $= 1.65$

7. continued
$$z = \frac{(\overline{X}_1 - \overline{X}_2) - (\mu_1 - \mu_2)}{\sqrt{\frac{s_1^2}{n_1} + \frac{s_2^2}{n_2}}} = \frac{(90 - 88) - 0}{\sqrt{\frac{5^2}{100} + \frac{6^2}{100}}} = 2.56$$

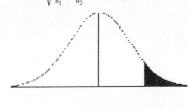

$0 \quad\quad 1.65 \quad \uparrow$
$\quad\quad\quad\quad\quad 2.56$

Reject the null hypothesis. There is enough evidence to support the claim that pulse rates of smokers are higher than the pulse rates of non-smokers.

9.
H_0: $\mu_1 \leq \mu_2$
H_1: $\mu_1 > \mu_2$ (claim)

C. V. $= 2.05$

$$z = \frac{(\overline{X}_1 - \overline{X}_2) - (\mu_1 - \mu_2)}{\sqrt{\frac{s_1^2}{n_1} + \frac{s_2^2}{n_2}}} = \frac{(61.2 - 59.4) - 0}{\sqrt{\frac{7.9^2}{84} + \frac{7.9^2}{34}}} = 1.12$$

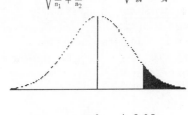

$0 \quad \uparrow 2.05$
$\quad\quad 1.12$

Do not reject the null hypothesis. There is not enough evidence to support the claim that noise levels in the corridors is higher than in the clinics.

11.
H_0: $\mu_1 \geq \mu_2$
H_1: $\mu_1 < \mu_2$ (claim)

C. V. $= -1.65$

$$z = \frac{(\overline{X}_1 - \overline{X}_2) - (\mu_1 - \mu_2)}{\sqrt{\frac{s_1^2}{n_1} + \frac{s_2^2}{n_2}}} = \frac{(3.16 - 3.28) - 0}{\sqrt{\frac{0.52^2}{103} + \frac{0.46^2}{225}}} = -2.01$$

11. continued

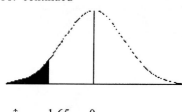

$\uparrow \quad -1.65 \quad 0$
-2.01

Reject the null hypothesis. There is enough evidence to support the claim that leavers have a lower GPA than stayers.

13.
$H_0: \mu_1 \le \mu_2$
$H_1: \mu_1 > \mu_2$ (claim)

C. V. = 2.33
$\overline{X}_1 = \$9224$ $\qquad$ $\overline{X}_2 = \$8497.5$
$s_1 = 3829.826$ $\qquad$ $s_2 = 2745.293$

$z = \dfrac{(\overline{X}_1 - \overline{X}_2) - (\mu_1 - \mu_2)}{\sqrt{\frac{s_1^2}{n_1} + \frac{s_2^2}{n_2}}}$

$z = \dfrac{(9224 - 8497.5) - 0}{\sqrt{\frac{3829.826^2}{50} + \frac{2745.293^2}{50}}} = 1.09$

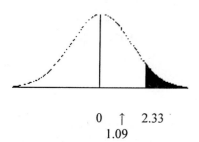

$0 \quad \uparrow \quad 2.33$
1.09

Do not reject the null hypothesis. There is not enough evidence to support the claim that colleges spent more money on men's sports than women's.

15.
$H_0: \mu_1 = \mu_2$
$H_1: \mu_1 \ne \mu_2$ (claim)

$z = \dfrac{(\overline{X}_1 - \overline{X}_2) - (\mu_1 - \mu_2)}{\sqrt{\frac{s_1^2}{n_1} + \frac{s_2^2}{n_2}}} = \dfrac{(3.05 - 2.96) - 0}{\sqrt{\frac{0.75^2}{103} + \frac{0.75^2}{225}}}$

$z = 1.01$
Area = 0.3438
P-value = 2(0.5 - 0.3438) = 0.3124

15. continued
Since P-value > 0.05, do not reject the null hypothesis. There is not enough evidence to support the claim that there is a difference in scores.

17.
$\overline{D} = 83.6 - 79.2 = 4.4$
$(\overline{X}_1 - \overline{X}_2) - z_{\frac{\alpha}{2}} \sqrt{\frac{\sigma_1^2}{n_1} + \frac{\sigma_2^2}{n_2}} < \mu_1 - \mu_2 <$
$\qquad (\overline{X}_1 - \overline{X}_2) + z_{\frac{\alpha}{2}} \sqrt{\frac{\sigma_1^2}{n_1} + \frac{\sigma_2^2}{n_2}}$

$4.4 - (1.65)\sqrt{\frac{4.3^2}{36} + \frac{3.8^2}{36}} < \mu_1 - \mu_2 <$
$\qquad 4.4 + (1.65)\sqrt{\frac{4.3^2}{36} + \frac{3.8^2}{36}}$

$2.8 < \mu_1 - \mu_2 < 6.0$

19.
$\overline{D} = 28.6 - 32.9 = -4.3$

$(\overline{X}_1 - \overline{X}_2) - z_{\frac{\alpha}{2}} \sqrt{\frac{\sigma_1^2}{n_1} + \frac{\sigma_2^2}{n_2}} < \mu_1 - \mu_2 <$
$\qquad (\overline{X}_1 - \overline{X}_2) + z_{\frac{\alpha}{2}} \sqrt{\frac{\sigma_1^2}{n_1} + \frac{\sigma_2^2}{n_2}}$

$-4.3 - (2.58)\sqrt{\frac{5.1^2}{30} + \frac{4.4^2}{40}} < \mu_1 - \mu_2 <$
$\qquad -4.3 + (2.58)\sqrt{\frac{5.2^2}{30} + \frac{4.4^2}{40}}$

$-7.3 < \mu_1 - \mu_2 < -1.3$

21.
$H_0: \mu_1 - \mu_2 \le 8$ (claim)
$H_1: \mu_1 - \mu_2 > 8$

C. V. = 1.65

$z = \dfrac{(\overline{X}_1 - \overline{X}_2) - K}{\sqrt{\frac{s_1^2}{n_1} + \frac{s_2^2}{n_2}}} = \dfrac{(110 - 104) - 8}{\sqrt{\frac{15^2}{60} + \frac{15^2}{60}}} = -0.73$

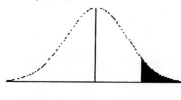

$\uparrow \quad 0 \qquad 1.65$
-0.73

Do not reject the null hypothesis. There is not enough evidence to reject the claim that private school students have exam scores

21. continued
that are at most 8 points higher than public
school students.

EXERCISE SET 9-3

1.
It should be the larger of the two variances.

3.
One d.f. is used for the variance associated
with the numerator and one is used for the
variance associated with the denominator.

5.
a. d. f. N = 15, d. f. D = 22; C. V. = 3.36
b. d. f. N = 24, d. f. D = 13; C. V. = 3.59
c. d. f. N = 45, d. f. D = 29; C. V. = 2.03
d. d. f. N = 20, d. f. D = 16; C. V. = 2.28
e. d. f. N = 10, d. f. D = 10; C. V. = 2.98

6.
Note: Specific P-values are in parentheses.
a. $0.025 < $ P-value $ < 0.05$ (0.033)
b. $0.05 < $ P-value $ < 0.10$ (0.072)
c. P-value $= 0.05$
d. $0.005 < $ P-value $ < 0.01$ (0.006)
e. P-value $= 0.05$
f. P > 0.10 (0.112)
g. $0.05 < $ P-value $ < 0.10$ (0.068)
h. $0.01 < $ P-value $ < 0.02$ (0.015)

7.
H_0: $\sigma_1^2 = \sigma_2^2$
H_1: $\sigma_1^2 \neq \sigma_2^2$ (claim)

C. V. = 2.53 $\alpha = \frac{0.10}{2}$
d. f. N = 14 d. f. D = 14
$F = \frac{s_1^2}{s_2^2} = \frac{13.12^2}{6.17^2} = 4.52$

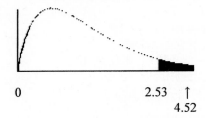

0 2.53 ↑
4.52

Reject the null hypothesis. There is enough
evidence to support the claim that there is a
difference in the variances of the best seller
lists for fiction and non-fiction.

9.
H_0: $\sigma_1^2 = \sigma_2^2$

H_1: $\sigma_1^2 \neq \sigma_2^2$ (claim)

$s_1 = 25.97$ $s_2 = 72.74$
C. V. = 2.86 $\alpha = \frac{0.05}{2}$
d. f. N = 15 d. f. D = 15

$F = \frac{s_1^2}{s_2^2} = \frac{72.74^2}{25.97^2} = 7.85$

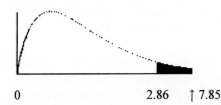

0 2.86 ↑ 7.85

Reject the null hypothesis. There is enough
evidence to support the claim that the
variances of the values of tax exempt
properties are different.

11.
H_0: $\sigma_1^2 = \sigma_2^2$
H_1: $\sigma_1^2 \neq \sigma_2^2$ (claim)

$s_1 = 33.99$ $s_2 = 33.99$
C. V. = 4.99 $\alpha = \frac{0.05}{2}$
d. f. N = 7 d. f. D = 7
$F = \frac{s_1^2}{s_2^2} = \frac{(33.99)^2}{(33.99)^2} = 1$

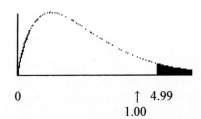

0 ↑ 4.99
1.00

Do not reject the null hypothesis. There is
not enough evidence to support the claim
that the variance in the number of calories
differs between the two brands.

13.
H_0: $\sigma_1^2 \leq \sigma_2^2$

H_1: $\sigma_1^2 > \sigma_2^2$ (claim)

$s_1 = 32$ $s_2 = 28$
$n_1 = 100$ $n_2 = 100$

13. continued
C. V. = 1.53 $\alpha = 0.05$
d. f. N = 99   d. f. D = 99

$F = \frac{s_1^2}{s_2^2} = \frac{(32)^2}{(28)^2} = 1.306$

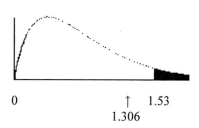

0   ↑   1.53
                  1.306

Reject the null hypothesis. There is enough evidence to support the claim that the variation of blood pressure of overweight individuals is greater than the variation of blood pressure of normal weight individuals.

15.
$H_0: \sigma_1^2 = \sigma_2^2$

$H_1: \sigma_1^2 \neq \sigma_2^2$ (claim)

Research: $s_1 = 5501.118$
Primary Care: $s_2 = 5238.809$

C. V. = 4.03 $\alpha = \frac{0.05}{2}$
d. f. N = 9   d. f. D = 9
$F = \frac{s_1^2}{s_2^2} = \frac{(5501.118)^2}{(5238.809)^2} = 1.10$

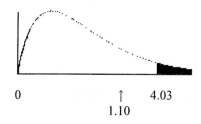

0   ↑   4.03
                  1.10

Do not reject the null hypothesis. There is not enough evidence to support the claim that there is a difference between the variances in tuition costs.

17.
$H_0: \sigma_1^2 = \sigma_2^2$ (claim)
$H_1: \sigma_1^2 \neq \sigma_2^2$

$s_1 = 130.496$ $s_2 = 73.215$
C. V. = 3.87 $\alpha = \frac{0.10}{2}$
d. f. N = 6   d. f. D = 7

17. continued
$F = \frac{s_1^2}{s_2^2} = \frac{(130.496)^2}{(73.215)^2} = 3.18$

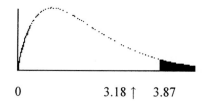

0   3.18 ↑   3.87

Do not reject the null hypothesis. There is not enough evidence to reject the claim that the variances of the heights are equal.

19.

Men	Women
$s_1^2 = 2.363$	$s_2^2 = 0.444$
$n_1 = 15$	$n_2 = 15$

$H_0: \sigma_1^2 = \sigma_2^2$ (claim)
$H_1: \sigma_1^2 \neq \sigma_2$

$\alpha = 0.05$ P-value = 0.004
d. f. N = 14   d. f. D = 14
$F = \frac{s_1^2}{s_2^2} = \frac{2.363}{0.444} = 5.32$

Since P-value < 0.05, reject the null hypothesis. There is enough evidence to reject the claim that the variances in weights are equal.

EXERCISE SET 9-4

1.
$H_0: \sigma_1^2 = \sigma_2^2$
$H_1: \sigma_1^2 \neq \sigma_2^2$
d. f. N = 9   d. f. D = 9 $\alpha = \frac{0.05}{2}$
$F = \frac{3256^2}{2341^2} = 1.93$ C. V. = 4.03
Do not reject. The variances are equal.

$H_0: \mu_1 = \mu_2$
$H_1: \mu_1 \neq \mu_2$ (claim)
C. V. = ± 2.101 d. f. = 18
$t = \frac{(\overline{X}_1 - \overline{X}_2) - (\mu_1 - \mu_2)}{\sqrt{\frac{(n_1-1)s_1^2 + (n_2-1)s_2^2}{n_1+n_2-2}}\sqrt{\frac{1}{n_1} + \frac{1}{n_2}}}$

$t = \frac{(83,256 - 88,354) - 0}{\sqrt{\frac{9(3256)^2 + 9(2341)^2}{18}}\sqrt{\frac{1}{10} + \frac{1}{10}}}$

$t = -4.02$

1. continued

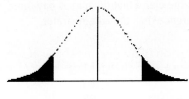

$\uparrow -2.101 \qquad 0 \qquad 2.101$
-4.02

Reject the null hypothesis. There is enough evidence to support the claim that there is a significant difference in the values of the homes based upon the appraisers' values.

Confidence Interval:

$$-5098 - 2.101\left(\sqrt{\frac{9(3256)^2 + 9(2341)^2}{18}}\right.$$

$$\left.\sqrt{\frac{1}{10} + \frac{1}{10}}\right) < \mu_1 - \mu_2 <$$

$$-5098 + 2.101\left(\sqrt{\frac{9(3256)^2 + 9(2341)^2}{18}}\right.$$

$$\left.\sqrt{\frac{1}{10} + \frac{1}{10}}\right) =$$

$$-5098 - 2.101(1268.14) < \mu_1 - \mu_2 <$$
$$-5098 + 2.101(1268.14)$$

$$-\$7762 < \mu_1 - \mu_2 < -\$2434$$

3.
H_0: $\sigma_1^2 = \sigma_2^2$
H_1: $\sigma_1^2 \neq \sigma_2^2$

d. f. N = 14   d. f. D = 14 $\alpha = \frac{0.05}{2}$

$F = \frac{20,000^2}{20,000^2} = 1$ C. V. = 3.05

Do not reject. The variances are equal.

H_0: $\mu_1 = \mu_2$
H_1: $\mu_1 \neq \mu_2$ (claim)

C. V. = ± 2.048
d. f. = 14 + 14 - 2 = 28

$$t = \frac{(\overline{X}_1 - \overline{X}_2) - (\mu_1 - \mu_2)}{\sqrt{\frac{(n_1-1)s_1^2 + (n_2-1)s_2^2}{n_1+n_2-2}}\sqrt{\frac{1}{n_1} + \frac{1}{n_2}}}$$

$$t = \frac{(501,580 - 513,360) - 0}{\sqrt{\frac{14(20,000^2) + 14(20,000^2)}{15 + 15 - 2}}\sqrt{\frac{1}{15} + \frac{1}{15}}}$$

$$t = -1.61$$

3. continued

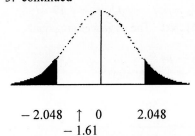

$-2.048 \quad \uparrow \quad 0 \qquad 2.048$
$\qquad -1.61$

Do not reject the null hypothesis. There is enough evidence to support the claim that there is no difference between the salaries.

5.
H_0: $\sigma_1^2 = \sigma_2^2$
H_1: $\sigma_1^2 \neq \sigma_2^2$
$\overline{X}_1 = 37.167$ $\overline{X}_2 = 25$
$s_1 = 13.2878$ $s_2 = 15.7734$
d. f. N = 5   d. f. D = 5 $\alpha = 0.01$
$F = \frac{15.7734^2}{13.2878^2} = 1.41$ C. V. = 14.94
Do not reject. The variances are equal.

H_0: $\mu_1 \leq \mu_2$
H_1: $\mu_1 > \mu_2$ (claim)

C. V. = 2.764   d. f. = 10

$$t = \frac{(\overline{X}_1 - \overline{X}_2) - (\mu_1 - \mu_2)}{\sqrt{\frac{(n_1-1)s_1^2 + (n_2-1)s_2^2}{n_1+n_2-2}}\sqrt{\frac{1}{n_1} + \frac{1}{n_2}}}$$

$$t = \frac{(37.167 - 25) - 0}{\sqrt{\frac{5(13.2878)^2 + 5(15.7734)^2}{6+6-2}}\sqrt{\frac{1}{6} + \frac{1}{6}}} = 1.45$$

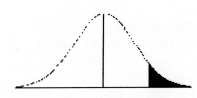

$0 \; 1.45 \uparrow 2.764$

Do not reject the null hypothesis. There is not enough evidence to support the claim that the average number of family day care centers is greater than the average number of day care centers.

7.
H_0: $\sigma_1^2 = \sigma_2^2$
H_1: $\sigma_1^2 \neq \sigma_2^2$
d. f. N = 9   d. f. D = 13 $\alpha = 0.02$

7. continued

$F = \frac{5.6^2}{4.3^2} = 1.7$ C. V. = 4.19

Do not reject. The variances are equal.

H_0: $\mu_1 = \mu_2$
H_1: $\mu_1 \neq \mu_2$ (claim)

C. V. = ± 2.508 d. f. = 22

$t = \frac{(\overline{X}_1 - \overline{X}_2) - (\mu_1 - \mu_2)}{\sqrt{\frac{(n_1 - 1)s_1^2 + (n_2 - 1)s_2^2}{n_1 + n_2 - 2}}\sqrt{\frac{1}{n_1} + \frac{1}{n_2}}}$

$t = \frac{(21 - 27) - 0}{\sqrt{\frac{9(5.6)^2 + 13(4.3)^2}{10 + 14 - 2}}\sqrt{\frac{1}{10} + \frac{1}{14}}} = -2.97$

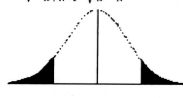

$\uparrow$ -2.508 0   2.508
-2.97

Reject the null hypothesis. There is enough evidence to support the claim that there is a difference in the average times of the two groups.

Confidence Interval:
$-6 - 2.508(2.02) < \mu_1 - \mu_2 <$
$\qquad\qquad\qquad -6 + 2.508(2.02)$
$-11.1 < \mu_1 - \mu_2 < -0.93$

9.
H_0: $\sigma_1^2 = \sigma_2^2$
H_1: $\sigma_1^2 \neq \sigma_2^2$
d. f. N = 26   d. f. D = 11 $\alpha = 0.05$
$F = \frac{5.75^2}{3^2} = 3.67$ P-value = 0.021

$0.02 <$ P-value < 0.05
Reject. The variances are unequal.

H_0: $\mu_1 \geq \mu_2$
H_1: $\mu_1 < \mu_2$ (claim)

P-value < 0.005 d. f. = 11

$t = \frac{(\overline{X}_1 - \overline{X}_2) - (\mu_1 - \mu_2)}{\sqrt{\frac{s_1^2}{n_1} + \frac{s_2^2}{n_2}}} = \frac{(56 - 63) - 0}{\sqrt{\frac{3^2}{12} + \frac{5.75^2}{27}}}$

$t = -4.98$

Since P-value < 0.05, reject the null hypothesis. There is enough evidence to

9. continued
support the claim that the nurses pay more for insurance than the administrators.

11.

White Mice	Brown Mice
$\overline{X}_1 = 17$	$\overline{X}_2 = 16.67$
$s_1 = 4.56$	$s_2 = 5.05$
$n_1 = 6$	$n_2 = 6$

H_0: $\sigma_1^2 = \sigma_2^2$
H_1: $\sigma_1^2 \neq \sigma_2^2$
d. f. N = 5   d. f. D = 5 $\alpha = \frac{0.05}{2}$
$F = \frac{5.05^2}{4.56^2} = 1.23$ C. V. = 7.15
Do not reject. The variances are equal.

H_0: $\mu_1 = \mu_2$
H_1: $\mu_1 \neq \mu_2$ (claim)

C. V. = ± 2.228 d. f. = 10

$t = \frac{(17 - 16.67) - 0}{\sqrt{\frac{5(4.56)^2 + 5(5.05)^2}{6 + 6 - 2}}\sqrt{\frac{1}{6} + \frac{1}{6}}} = 0.119$

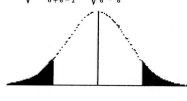

-2.228 0 $\uparrow$ 2.228
$\qquad\qquad$ 0.119

Do not reject the null hypothesis. There is not enough evidence to support the claim that the color of the mice made a difference.

Confidence Interval:
$0.33 - 2.228(2.78) < \mu_1 - \mu_2 <$
$\qquad\qquad\qquad 0.33 + 2.228(2.78)$
$-5.9 < \mu_1 - \mu_2 < 6.5$

13.
Private: $\overline{X} = \$16{,}147.5$ s = 4023.7
Public: $\overline{X} = \$9039.9$ s = 3325.5

F test:
d. f. N = 6 - 1 = 5
d. f. D = 7 - 1 = 6
C. V. = 5.99
$F = \frac{4023.7^2}{3325.5^2} = 1.46$
Do not reject. The variances are equal.

13. continued
Confidence Interval:

$$t_{\frac{\alpha}{2}}\sqrt{\frac{(n_1-1)s_1^2+(n_2-1)s_2^2}{n_1+n_2-2}}\sqrt{\frac{1}{n_1}+\frac{1}{n_2}}=$$

$$2.201\sqrt{\frac{5(4023.7)^2+6(3325.5)^2}{6+7-2}}\sqrt{\frac{1}{6}+\frac{1}{7}}$$

$$=4481.04$$

$$16,147.5-9039.9=7107.6$$

$$7107.6-4481.04<\mu_1-\mu_2<$$
$$7107.6+4481.04$$

$$\$2626.60<\mu_1-\mu_2<\$11,588.64$$

EXERCISE SET 9-5

1.
a. dependent
b. dependent
c. independent
d. dependent
e. independent

3.

Before	After	D	D^2
9	9	0	0
12	17	-5	25
6	9	-3	9
15	20	-5	25
3	2	1	1
18	21	-3	9
10	15	-5	25
13	22	-9	81
7	6	1	1
		$\sum D=-28$	$\sum D^2=176$

H_0: $\mu_D \geq 0$
H_1: $\mu_D < 0$ (claim)

C. V. $= -1.397$ d. f. $= 8$

$$\overline{D}=\frac{\sum D}{n}=-3.11$$

$$s_D=\sqrt{\frac{\sum D^2-\frac{(\sum D)^2}{n}}{n-1}}=\sqrt{\frac{176-\frac{(-28)^2}{9}}{8}}=3.33$$

$$t=\frac{-3.11-0}{\frac{3.33}{\sqrt{9}}}=-2.8$$

3. continued

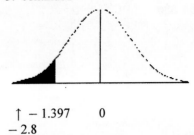

$\uparrow -1.397$ 0
-2.8

Reject the null hypothesis. There is enough evidence to support the claim that the seminar increased the number of hours students studied.

5.

F - S	S - Th	D	D^2
4	8	-4	16
7	5.5	1.5	2.25
10.5	7.5	3	9
12	8	4	16
11	7	4	16
9	6	3	9
6	6	0	0
9	8	1	1
		$\sum D=12.5$	$\sum D^2=69.25$

H_0: $\mu_D = 0$
H_1: $\mu_D \neq 0$ (claim)

C. V. $= \pm 2.365$ d. f. $= 7$
$$\overline{D}=\frac{\sum D}{n}=\frac{12.5}{8}=1.5625$$

$$s_D=\sqrt{\frac{\sum D^2-\frac{(\sum D)^2}{n}}{n-1}}$$

$$=\sqrt{\frac{69.25-\frac{(12.5)^2}{8}}{7}}=2.665$$

$$t=\frac{1.5625-0}{\frac{2.665}{\sqrt{8}}}=1.6583$$

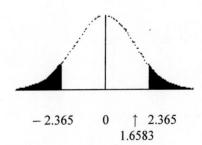

-2.365 0 $\uparrow$ 2.365
1.6583

Do not reject the null hypothesis. There is not enough evidence to support the claim that there is a difference in the mean number of hours slept.

7.

Before	After	D	D^2
12	9	3	9
9	6	3	9
0	1	-1	1
5	3	2	4
4	2	2	4
3	3	0	0

$$\sum D = 9 \quad \sum D^2 = 27$$

H_0: $\mu_D \leq 0$
H_1: $\mu_D > 0$ (claim)

C. V. $= 2.571$ d. f. $= 5$

$$\overline{D} = \frac{\sum D}{n} = \frac{9}{6} = 1.5$$

$$s_D = \sqrt{\frac{\sum D^2 - \frac{(\sum D)^2}{n}}{n-1}} = \sqrt{\frac{27 - \frac{9^2}{6}}{5}} = 1.64$$

$$t = \frac{1.5 - 0}{\frac{1.64}{\sqrt{6}}} = 2.24$$

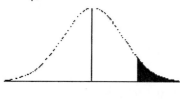

$$0 \quad \uparrow \quad 2.571$$
$$2.24$$

Do not reject the null hypothesis. There is not enough evidence to support the claim that the errors have been reduced.

9.

A	B	D	D^2
87	83	4	16
92	95	-3	9
78	79	-1	1
83	83	0	0
88	86	2	4
90	93	-3	9
84	80	4	16
93	86	7	49

$$\sum D = 10 \quad \sum D^2 = 104$$

H_0: $\mu_D = 0$
H_1: $\mu_D \neq 0$ (claim)

P-value $= 0.361$ d. f. $= 7$

$$\overline{D} = \frac{\sum D}{n} = \frac{10}{8} = 1.25$$

$$s_D = \sqrt{\frac{\sum D^2 - \frac{(\sum D)^2}{n}}{n-1}} = \sqrt{\frac{104 - \frac{10^2}{8}}{7}} = 3.62$$

9. continued

$$t = \frac{1.25 - 0}{\frac{3.62}{\sqrt{8}}} = 0.978$$

$0.20 <$ P-value < 0.50 Do not reject the null hypothesis since P-value > 0.01. There is not enough evidence to support the claim that there is a difference in the pulse rates.

Confidence Interval:
$$1.25 - 3.499\left(\frac{3.62}{\sqrt{8}}\right) < \mu_D <$$
$$1.25 + 3.499\left(\frac{3.62}{\sqrt{8}}\right)$$
$$-3.23 < \mu_D < 5.73$$

11.
Using the previous problem, $\overline{D} = -1.5625$ whereas the mean of the 1994 values is 95.375 and the mean of the 1999 values is 96.9375; hence,
$$\overline{D} = 95.375 - 96.9375 = -1.5625$$

EXERCISE SET 9-6

1A.
Use $\hat{p} = \frac{X}{n}$ and $\hat{q} = 1 - \hat{p}$

a. $\hat{p} = \frac{34}{48}$ $\hat{q} = \frac{14}{48}$

b. $\hat{p} = \frac{28}{75}$ $\hat{q} = \frac{47}{75}$

c. $\hat{p} = \frac{50}{100}$ $\hat{q} = \frac{50}{100}$

d. $\hat{p} = \frac{6}{24}$ $\hat{q} = \frac{18}{24}$

e. $\hat{p} = \frac{12}{144}$ $\hat{q} = \frac{132}{144}$

1B.
a. x $= 0.16(100) = 16$
b. x $= 0.08(50) = 4$
c. x $= 0.06(80) = 4.8$
d. x $= 0.52(200) = 104$
e. x $= 0.20(150) = 30$

3.
$$\hat{p}_1 = \frac{X_1}{n_1} = \frac{80}{150} = 0.533 \quad \hat{p}_2 = \frac{30}{100} = 0.3$$

$$\overline{p} = \frac{X_1 + X_2}{n_1 + n_2} = \frac{80 + 30}{150 + 100} = \frac{110}{250} = 0.44$$

$$\overline{q} = 1 - \overline{p} = 1 - 0.44 = 0.56$$

H_0: $p_1 = p_2$
H_1: $p_1 \neq p_2$ (claim)

3. continued

C. V. $= \pm 1.96$

$$z = \frac{(\hat{p}_1 - \hat{p}_2) - (p_1 - p_2)}{\sqrt{(\bar{p})(\bar{q})(\frac{1}{n_1} + \frac{1}{n_2})}} = \frac{(0.533 - 0.3) - 0}{\sqrt{(0.44)(0.56)(\frac{1}{150} + \frac{1}{100})}}$$

$z = 3.64$

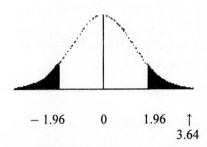

$$-1.96 \qquad 0 \qquad 1.96 \quad \uparrow$$
$$3.64$$

Reject the null hypothesis. There is enough evidence to support the claim that there is a significant difference in the proportions.

5.
$$\hat{p}_1 = \frac{X_1}{n_1} = \frac{112}{150} = 0.7467 \quad \hat{p}_2 = \frac{150}{200} = 0.75$$

$$\bar{p} = \frac{X_1 + X_2}{n_1 + n_2} = \frac{112 + 150}{150 + 200} = 0.749$$

$$\bar{q} = 1 - \bar{p} = 1 - 0.749 = 0.251$$

H_0: $p_1 = p_2$
H_1: $p_1 \neq p_2$ (claim)

C. V. $= \pm 1.96$

$$z = \frac{(\hat{p}_1 - \hat{p}_2) - (p_1 - p_2)}{\sqrt{(\bar{p})(\bar{q})(\frac{1}{n_1} + \frac{1}{n_2})}} = \frac{(0.7467 - 0.75) - 0}{\sqrt{(0.749)(0.251)(\frac{1}{150} + \frac{1}{200})}}$$

$z = -0.07$

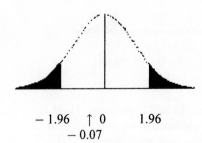

$$-1.96 \quad \uparrow 0 \qquad 1.96$$
$$-0.07$$

Do not reject the null hypothesis. There is not enough evidence to support the claim that the proportions are different.

7.
$\hat{p}_1 = 0.83 \qquad \hat{p}_2 = 0.75$
$X_1 = 0.83(100) = 83$
$X_2 = 0.75(100) = 75$

$$\bar{p} = \frac{83 + 75}{100 + 100} = 0.79 \quad \bar{q} = 1 - 0.79 = 0.21$$

H_0: $p_1 = p_2$ (claim)
H_1: $p_1 \neq p_2$
C. V. $= \pm 1.96 \quad \alpha = 0.05$

$$z = \frac{(\hat{p}_1 - \hat{p}_2) - (p_1 - p_2)}{\sqrt{(\bar{p})(\bar{q})(\frac{1}{n_1} + \frac{1}{n_2})}} = \frac{(0.83 - 0.75) - 0}{\sqrt{(0.79)(0.21)(\frac{1}{100} + \frac{1}{100})}}$$

$z = 1.39$

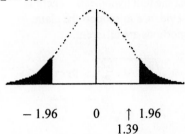

$$-1.96 \qquad 0 \quad \uparrow 1.96$$
$$1.39$$

Do not reject the null hypothesis. There is not enough evidence to reject the claim that the proportions are equal.

$$(\hat{p}_1 - \hat{p}_2) - z_{\frac{\alpha}{2}} \sqrt{\frac{\hat{p}_1\hat{q}_1}{n_1} + \frac{\hat{p}_2\hat{q}_2}{n_2}} < p_1 - p_2 <$$

$$(\hat{p}_1 - \hat{p}_2) + z_{\frac{\alpha}{2}} \sqrt{\frac{\hat{p}_1\hat{q}_1}{n_1} + \frac{\hat{p}_2\hat{q}_2}{n_2}}$$

$$0.08 - 1.96\sqrt{\frac{0.83(0.17)}{100} + \frac{0.75(0.25)}{100}} < p_1 - p_2$$

$$< 0.08 + 1.96\sqrt{\frac{0.83(0.17)}{100} + \frac{0.75(0.25)}{100}}$$

$$-0.032 < p_1 - p_2 < 0.192$$

9.
$\hat{p}_1 = 0.55 \qquad \hat{p}_2 = 0.45$

$X_1 = 0.55(80) = 44 \quad X_2 = 0.45(90) = 40.5$

$$\bar{p} = \frac{X_1 + X_2}{n_1 + n_2} = \frac{44 + 40.5}{80 + 90} = 0.497$$

$$\bar{q} = 1 - \bar{p} = 1 - 0.497 = 0.503$$

H_0: $p_1 = p_2$
H_1: $p_1 \neq p_2$ (claim)

C. V. $= \pm 2.58 \quad \alpha = 0.01$

9. continued

$$z = \frac{(\hat{p}_1 - \hat{p}_2) - (p_1 - p_2)}{\sqrt{(\bar{p})(\bar{q})(\frac{1}{n_1} + \frac{1}{n_2})}} = \frac{(0.55 - 0.45) - 0}{\sqrt{(0.497)(0.503)(\frac{1}{80} + \frac{1}{90})}}$$

$z = 1.302$

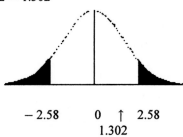

$-2.58 \qquad 0 \quad \uparrow \quad 2.58$

1.302

Do not reject the null hypothesis. There is not enough evidence to support the claim that the proportions are different.

$$(\hat{p}_1 - \hat{p}_2) - z_{\frac{\alpha}{2}}\sqrt{\frac{\hat{p}_1\hat{q}_1}{n_1} + \frac{\hat{p}_2\hat{q}_2}{n_2}} < p_1 - p_2 <$$

$$(\hat{p}_1 - \hat{p}_2) + z_{\frac{\alpha}{2}}\sqrt{\frac{\hat{p}_1\hat{q}_1}{n_1} + \frac{\hat{p}_2\hat{q}_2}{n_2}}$$

$$0.1 - 2.58\sqrt{\frac{0.55(0.45)}{80} + \frac{0.45(0.55)}{90}} < p_1 - p_2$$

$$< 0.1 + 2.58\sqrt{\frac{0.55(0.45)}{80} + \frac{0.45(0.55)}{90}}$$

$$-0.097 < p_1 - p_2 < 0.297$$

11.

$\hat{p}_1 = \frac{45}{80} = 0.5625 \qquad \hat{p}_2 = \frac{63}{120} = 0.525$

$\bar{p} = \frac{X_1 + X_2}{n_1 + n_2} = \frac{45 + 63}{80 + 120} = 0.54$

$\bar{q} = 1 - \bar{p} = 1 - 0.54 = 0.46$

H_0: $p_1 = p_2$
H_1: $p_1 \neq p_2$ (claim)

C. V. $= \pm 1.96 \qquad \alpha = 0.05$

$$z = \frac{(\hat{p}_1 - \hat{p}_2) - (p_1 - p_2)}{\sqrt{(\bar{p})(\bar{q})(\frac{1}{n_1} + \frac{1}{n_2})}} = \frac{(0.5625 - 0.525) - 0}{\sqrt{(0.54)(0.46)(\frac{1}{80} + \frac{1}{120})}}$$

$z = 0.521$

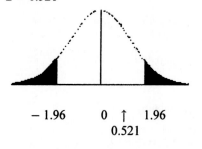

$-1.96 \qquad 0 \quad \uparrow \quad 1.96$

0.521

11. continued

Do not reject the null hypothesis. There is not enough evidence to support the claim that there is a difference in the proportions.

$$(\hat{p}_1 - \hat{p}_2) - z_{\frac{\alpha}{2}}\sqrt{\frac{\hat{p}_1\hat{q}_1}{n_1} + \frac{\hat{p}_2\hat{q}_2}{n_2}} < p_1 - p_2 <$$

$$(\hat{p}_1 - \hat{p}_2) + z_{\frac{\alpha}{2}}\sqrt{\frac{\hat{p}_1\hat{q}_1}{n_1} + \frac{\hat{p}_2\hat{q}_2}{n_2}}$$

$$0.0375 - 1.96\sqrt{\frac{0.5625(0.4375)}{80} + \frac{0.525(0.475)}{120}} <$$

$$p_1 - p_2 < 0.0375 + 1.96\sqrt{\frac{0.5625(0.4375)}{80} + \frac{0.525(0.475)}{120}}$$

$$-0.103 < p_1 - p_2 < 0.178$$

13.

$\hat{p}_1 = \frac{X_1}{n_1} = \frac{50}{200} = 0.25$

$\hat{p}_2 = \frac{X_2}{n_2} = \frac{93}{300} = 0.31$

$\bar{p} = \frac{X_1 + X_2}{n_1 + n_2} = \frac{50 + 93}{200 + 300} = 0.286$

$\bar{q} = 1 - \bar{p} = 1 - 0.286 = 0.714$

H_0: $p_1 = p_2$
H_1: $p_1 \neq p_2$ (claim)

C. V. $= \pm 2.58 \quad \alpha = 0.01$

$$z = \frac{(\hat{p}_1 - \hat{p}_2) - (p_1 - p_2)}{\sqrt{(\bar{p})(\bar{q})(\frac{1}{n_1} + \frac{1}{n_2})}} = \frac{(0.25 - 0.31) - 0}{\sqrt{(0.286)(0.714)(\frac{1}{200} + \frac{1}{300})}}$$

$z = -1.45$

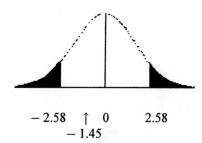

$-2.58 \quad \uparrow \quad 0 \qquad 2.58$

-1.45

Do not reject the null hypothesis. There is not enough evidence to support the claim that the proportions are different.

$$(\hat{p}_1 - \hat{p}_2) - z_{\frac{\alpha}{2}}\sqrt{\frac{\hat{p}_1\hat{q}_1}{n_1} + \frac{\hat{p}_2\hat{q}_2}{n_2}} < p_1 - p_2 <$$

$$(\hat{p}_1 - \hat{p}_2) + z_{\frac{\alpha}{2}}\sqrt{\frac{\hat{p}_1\hat{q}_1}{n_1} + \frac{\hat{p}_2\hat{q}_2}{n_2}}$$

13. continued

$$-0.06 - 2.58\sqrt{\frac{0.25(0.75)}{200} + \frac{0.31(0.69)}{300}} <$$

$$p_1 - p_2 < -0.06 + 2.58\sqrt{\frac{0.25(0.75)}{200} + \frac{0.31(0.69)}{300}}$$

$$-0.165 < p_1 - p_2 < 0.045$$

15.
$\alpha = 0.01$
$\hat{p}_1 = 0.8 \qquad \hat{q}_1 = 0.2$
$\hat{p}_2 = 0.6 \qquad \hat{q}_2 = 0.4$

$$\hat{p}_1 - \hat{p}_2 = 0.8 - 0.6 = 0.2$$

$$(\hat{p}_1 - \hat{p}_2) - z_{\frac{\alpha}{2}}\sqrt{\frac{\hat{p}_1\hat{q}_1}{n_1} + \frac{\hat{p}_2\hat{q}_2}{n_2}} < p_1 - p_2 <$$

$$(\hat{p}_1 - \hat{p}_2) + z_{\frac{\alpha}{2}}\sqrt{\frac{\hat{p}_1\hat{q}_1}{n_1} + \frac{\hat{p}_2\hat{q}_2}{n_2}}$$

$$0.2 - 2.58\sqrt{\frac{(0.8)(0.2)}{150} + \frac{(0.6)(0.4)}{200}} < p_1 - p_2 <$$

$$0.2 + 2.58\sqrt{\frac{(0.8)(0.2)}{150} + \frac{(0.6)(0.4)}{200}}$$

$$0.077 < p_1 - p_2 < 0.323$$

17.
$$\hat{p}_1 = \frac{X_1}{n_1} = \frac{43}{100} = 0.43 \quad \hat{p}_2 = \frac{58}{100} = 0.58$$

$$\bar{p} = \frac{X_1 + X_2}{n_1 + n_2} = \frac{43 + 58}{100 + 100} = 0.505$$

$$\bar{q} = 1 - \bar{p} = 1 - 0.505 = 0.495$$

H_0: $p_1 = p_2$
H_1: $p_1 \neq p_2$ (claim)

C. V. $= \pm 1.96$

$$z = \frac{(\hat{p}_1 - \hat{p}_2) - (p_1 - p_2)}{\sqrt{(\bar{p})(\bar{q})(\frac{1}{n_1} + \frac{1}{n_2})}} = \frac{(0.43 - 0.58) - 0}{\sqrt{(0.505)(0.495)(\frac{1}{100} + \frac{1}{100})}}$$
$$z = -2.12$$

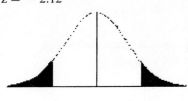

$\uparrow -1.96 \qquad 0 \qquad 1.96$
-2.12

17. continued
Reject the null hypothesis. There is enough evidence to support the claim that the proportions are different.

19.
$\hat{p}_1 = 0.2875 \qquad \hat{q}_1 = 0.7125$
$\hat{p}_2 = 0.2857 \qquad \hat{q}_2 = 0.7143$
$\hat{p}_1 - \hat{p}_2 = 0.0018$

$$(\hat{p}_1 - \hat{p}_2) - z_{\frac{\alpha}{2}}\sqrt{\frac{\hat{p}_1\hat{q}_1}{n_1} + \frac{\hat{p}_2\hat{q}_2}{n_2}} < p_1 - p_2 <$$

$$(\hat{p}_1 - \hat{p}_2) + z_{\frac{\alpha}{2}}\sqrt{\frac{\hat{p}_1\hat{q}_1}{n_1} + \frac{\hat{p}_2\hat{q}_2}{n_2}}$$

$$0.0018 - 1.96\sqrt{\frac{(0.2875)(0.7125)}{400} + \frac{(0.2857)(0.7143)}{350}} < p_1 - p_2$$

$$< 0.0018 + 1.96\sqrt{\frac{(0.2875)(0.7125)}{400} + \frac{(0.2857)(0.7143)}{350}}$$

$$-0.0631 < p_1 - p_2 < 0.0667$$

REVIEW EXERCISES - CHAPTER 9

1.
H_0: $\mu_1 \leq \mu_2$
H_1: $\mu_1 > \mu_2$ (claim)

CV $= 2.33 \qquad \alpha = 0.01$
$\bar{X}_1 = 120.1 \qquad \bar{X}_2 = 117.8$
$s_1 = 16.722 \qquad s_2 = 16.053$

$$z = \frac{(\bar{X}_1 - \bar{X}_2) - (\mu_1 - \mu_2)}{\sqrt{\frac{s_1^2}{n_1} + \frac{s_2^2}{n_2}}} = \frac{(120.1 - 117.8) - 0}{\sqrt{\frac{16.722^2}{36} + \frac{16.053^2}{35}}}$$

$z = 0.587$ or 0.59

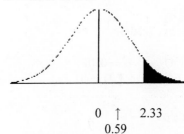

$0 \quad \uparrow \quad 2.33$
$\quad 0.59$

Do not reject the null hypothesis. There is not enough evidence to support the claim that single people do more pleasure driving than married people.

3.
H_0: $\sigma_1 = \sigma_2$
H_1: $\sigma_1 \neq \sigma_2$ (claim)

3. continued

C. V. = 2.77 $\alpha = 0.10$

d. f. N = 23   d. f. D = 10

$$F = \frac{13.2^2}{4.1^2} = 10.365$$

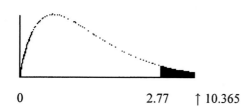

0   2.77   ↑ 10.365

Reject the null hypothesis. There is enough evidence to support the claim that there is a difference in standard deviations.

5.

H_0: $\sigma_1^2 \leq \sigma_2^2$

H_1: $\sigma_1^2 > \sigma_2^2$ (claim)

$\alpha = 0.05$

d. f. N = 9   d. f. D = 9

$F = \frac{s_1^2}{s_2^2} = \frac{6.3^2}{2.8^2} = 5.06$

The P-value for the F test is $0.01 < $ P-value < 0.025 (0.012). Since P-value < 0.05, reject the null hypothesis. There is enough evidence to support the claim that the variance of the number of speeding tickets on Route 19 is greater than the variance of the number of speeding tickets issued on Route 22.

7.

H_0: $\sigma_1^2 \leq \sigma_2^2$

H_1: $\sigma_1^2 > \sigma_2^2$ (claim)

C. V. = 1.47 $\alpha = 0.10$

d. f. N = 64   d. f. D = 41

$F = \frac{s_1^2}{s_2^2} = \frac{3.2^2}{2.1^2} = 2.32$

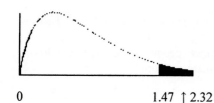

0   1.47 ↑ 2.32

Reject the null hypothesis. There is enough evidence to support the claim that the

7. continued

variation in the number of days factory workers miss per year due to illness is greater than the variation in the number of days hospital workers miss per year.

9.

H_0: $\sigma_1^2 = \sigma_2^2$

H_1: $\sigma_1^2 \neq \sigma_2^2$

$\overline{X}_1 = 72.9$ $\overline{X}_2 = 70.8$

$s_1 = 5.5$ $s_2 = 5.8$

CV = 1.98 $\alpha = 0.01$

dfN = 24   dfD = 24

$F = \frac{5.8^2}{5.5^2} = 1.11$

Do not reject H_0. The variances are equal.

H_0: $\mu_1 \leq \mu_2$

H_1: $\mu_1 > \mu_2$ (claim)

C. V. = 1.28   d. f. = 48 $\alpha = 0.10$

$$t = \frac{(\overline{X}_1 - \overline{X}_2) - (\mu_1 - \mu_2)}{\sqrt{\frac{(n_1 - 1)s_1^2 + (n_2 - 1)s_2^2}{n_1 + n_2 - 2}}\sqrt{\frac{1}{n_1} + \frac{1}{n_2}}}$$

$$t = \frac{(72.9 - 70.8) - 0}{\sqrt{\frac{24(5.5)^2 + 24(5.8)^2}{25 + 25 - 2}}\sqrt{\frac{1}{25} + \frac{1}{25}}} = 1.31$$

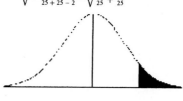

0   1.28 ↑ 1.31

Reject the null hypothesis. There is enough evidence to support the claim that it is warmer in Birmingham.

11.

H_0: $\sigma_1^2 = \sigma_2^2$

H_1: $\sigma_1^2 \neq \sigma_2^2$

$\alpha = 0.05$

dfN = 15   dfD = 11

$F = \frac{8256^2}{1311^2} = 39.66$

Reject H_0 since P-value < 0.05. The variances are unequal.

H_0: $\mu_1 \leq \mu_2$

H_1: $\mu_1 > \mu_2$ (claim)

$\alpha = 0.05$ P-value < 0.005

11. continued

$$t = \frac{(54{,}356 - 46{,}512) - 0}{\sqrt{\frac{8256^2}{16} + \frac{1311^2}{12}}} = 3.74$$

Reject the null hypothesis since P-value < 0.05. There is enough evidence to support the claim that incomes of the city residents are greater than the incomes of the suburban residents.

13.

Before	After	D	D^2
6	10	-4	16
8	12	-4	16
10	9	1	1
9	12	-3	9
5	8	-3	9
12	13	-1	1
9	8	1	1
7	10	-3	9

$$\sum D = -16 \qquad \sum D^2 = 62$$

H_0: $\mu_D \geq 0$
H_1: $\mu_D < 0$ (claim)

C. V. $= -1.895$ d. f. $= 7$ $\alpha = 0.05$

$$\overline{D} = \frac{\sum D}{n} = \frac{-16}{8} = -2$$

$$s_D = \sqrt{\frac{62 - \frac{(-16)^2}{8}}{7}} = 2.07$$

$$t = \frac{-2 - 0}{\frac{2.07}{\sqrt{8}}} = -2.73$$

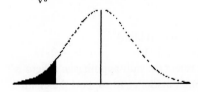

$\uparrow \quad -1.895 \qquad 0$
-2.73

Reject the null hypothesis. There is enough evidence to support the claim that the music has increased production.

15.

$$\hat{p}_1 = \frac{32}{50} = 0.64 \qquad\qquad \hat{p}_2 = \frac{24}{60} = 0.40$$

$$\overline{p} = \frac{32 + 24}{50 + 60} = 0.509$$

$$\overline{q} = 1 - 0.509 = 0.491$$

H_0: $p_1 = p_2$ (claim)
H_1: $p_1 \neq p_2$

15. continued

C. V. $= \pm 1.96$ $\alpha = 0.05$

$$z = \frac{(0.64 - 0.40) - 0}{\sqrt{(0.509)(0.491)(\frac{1}{50} + \frac{1}{60})}} = 2.51$$

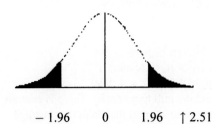

$-1.96 \qquad 0 \qquad 1.96 \quad \uparrow 2.51$

Reject the null hypothesis. There is enough evidence to reject the claim that the proportions are equal.

For the 95% confidence interval:
$\hat{p}_1 = 0.64 \qquad \hat{q}_1 = 0.36$
$\hat{p}_2 = 0.4 \qquad \hat{q}_2 = 0.6$

$$(0.64 - 0.4) - 1.96\sqrt{\frac{(0.64)(0.36)}{50} + \frac{(0.4)(0.6)}{60}}$$

$$< p_1 - p_2 < (0.64 - 0.4) + 1.96\sqrt{\frac{(0.64)(0.36)}{50} + \frac{(0.4)(0.6)}{60}}$$

$$0.24 - 1.96(0.0928) < p_1 - p_2 <$$

$$0.24 + 1.96(0.0928)$$

$$0.058 < p_1 - p_2 < 0.422$$

CHAPTER 9 QUIZ

1. False, there are different formulas for independent and dependent samples.
2. False, the samples are independent.
3. True
4. False, they can be right, left, or two tailed.
5. d.
6. a.
7. c.
8. b.
9. $\mu_1 = \mu_2$
10. pooled
11. normal
12. negative
13. $\frac{s_1^2}{s_2^2}$

14. H_0: $\mu_1 = \mu_2$
H_1: $\mu_1 \neq \mu_2$ (claim)
C. V. $= \pm 2.58$ $z = -3.69$

14. continued
Reject the null hypothesis. There is enough evidence to support the claim that there is a difference in the cholesterol levels of the two groups.
99% Confidence Interval:
$-10.2 < \mu_1 - \mu_2 < -1.8$

15. H_0: $\mu_1 \leq \mu_2$
H_1: $\mu_1 > \mu_2$ (claim)
C. V. = 1.28   z = 1.60
Reject the null hypothesis. There is enough evidence to support the claim that average rental fees for the east apartments is greater than the average rental fees for the west apartments.

16. H_0: $\sigma_1^2 = \sigma_2^2$
H_1: $\sigma_1^2 \neq \sigma_1^2$ (claim)
F = 1.637   P-value > 0.20 (0.357)
Do not reject the null hypothesis since P-value > 0.05. There is not enough evidence to support the claim that the variances are different.

17. H_0: $\sigma_1^2 = \sigma_2^2$
H_1: $\sigma_1^2 \neq \sigma_2^2$ (claim)
C. V. = 1.90   F = 1.296
Reject the null hypothesis. There is enough evidence to support the claim that the variances are different.

18. H_0: $\sigma_1^2 = \sigma_2^2$ (claim)
H_1 $\sigma_1^2 \neq \sigma_2^2$
C. V. = 3.53   F = 1.13
Do not reject the null hypothesis. There is not enough evidence to reject the claim that the standard deviations or the number of hours of television viewing are the same.

19. H_0: $\sigma_1^2 = \sigma_2^2$
H_1 $\sigma_1^2 \neq \sigma_2^2$ (claim)
C. V. = 3.01   F = 1.94
Do not reject the null hypothesis. There is enough evidence to support the claim that the variances are different.

20. H_0: $\sigma_1^2 \leq \sigma_2^2$
H_1 $\sigma_1^2 > \sigma_2^2$ (claim)
C. V. = 1.44   F = 1.474
Reject the null hypothesis. There is enough evidence to support the claim that the variance of days missed by teachers is

20. continued
greater than the variance of days missed by nurses.

21. H_0: $\sigma_1 = \sigma_2$
H_1 $\sigma_1 \neq \sigma_2$ (claim)
C. V. = 2.46   F = 1.65
Do not reject the null hypothesis. There is not enough evidence to support the claim that the standard deviations are different.

22. H_0: $\sigma_1^2 = \sigma_2^2$
H_1 $\sigma_1^2 \neq \sigma_2^2$
C. V. = 5.05   F = 1.23
Do not reject. The variances are equal.

H_0: $\mu_1 = \mu_2$
H_1: $\mu_1 \neq \mu_2$ (claim)
C. V. = ±2.779   t = 10.92

Reject the null hypothesis. There is enough evidence to support the claim that the average prices are different.

99% Confidence Interval:
$0.298 < \mu_1 - \mu_2 < 0.502$

23. H_0: $\sigma_1^2 = \sigma_2^2$
H_1 $\sigma_1^2 \neq \sigma_2^2$
C. V. = 9.6   F = 12.41
Reject. The variances are not equal.

H_0: $\mu_1 \geq \mu_2$
H_1: $\mu_1 < \mu_2$ (claim)
C. V. = -2.131 t = -2.07
Do not reject the null hypothesis. There is not enough evidence to support the claim that accidents have increased.

24. H_0: $\sigma_1^2 = \sigma_2^2$
H_1 $\sigma_1^2 \neq \sigma_2^2$
C. V. = 4.02   F = 6.155
Reject. The variances are unequal.

H_0: $\mu_1 = \mu_2$
H_1: $\mu_1 \neq \mu_2$ (claim)
C. V. = ±2.718   t = 9.807
Reject the null hypothesis. There is enough evidence to support the claim that the salaries are different.

98% Confidence Interval:
$\$6652 < \mu_1 - \mu_2 < \$11,757$

25. H_0: $\sigma_1^2 = \sigma_2^2$
H_1 $\sigma_1^2 \neq \sigma_2^2$
$F = 23.08$ P-value < 0.05
Reject. The variances are unequal.

H_0: $\mu_1 \leq \mu_2$
H_1: $\mu_1 > \mu_2$ (claim)
$t = 0.874$ $0.10 <$ P-value < 0.25 (0.198)
Do not reject the null hypothesis since P-value > 0.05. There is not enough evidence to support the claim that incomes of city residents is greater than incomes of rural residents.

26. H_0: $\mu_1 \geq \mu_2$
H_1: $\mu_1 < \mu_2$ (claim)
$\overline{D} = -6.5$ $s_D = 4.93$
C. V. $= -2.821$ $t = -4.17$
Reject the null hypothesis. There is enough evidence to support the claim that the sessions improved math skills.

27. H_0: $\mu_1 \geq \mu_2$
H_1: $\mu_1 < \mu_2$ (claim)
$\overline{D} = -0.8$ $s_D = 1.48$
C. V. $= -1.833$ $t = -1.71$
Do not reject the null hypothesis. There is not enough evidence to support the claim that egg production increased.

28. H_0: $p_1 = p_2$
H_1: $p_1 \neq p_2$ (claim)
C. V. $= \pm 1.65$ $z = -0.69$
Do not reject the null hypothesis. There is not enough evidence to support the claim that the proportions are different.

90% Confidence Interval:
$-0.101 < p_1 - p_2 < 0.041$

29. H_0: $p_1 = p_2$ (claim)
H_1: $p_1 \neq p_2$
C. V. $= \pm 1.96$ $z = 2.58$
Reject the null hypothesis. There is enough evidence to reject the claim that the proportions are equal.

95% Confidence Interval:
$0.067 < p_1 - p_2 < 0.445$

Note: Graphs are not to scale and are intended to convey a general idea.

Answers may vary due to rounding.

EXERCISE SET 10-3

1.
Two variables are related when there exists a discernible pattern between them.

3.
r, ρ (rho)

5.
A positive relationship means that as x increases, y also increases.
A negative relationship means that as x increases, y decreases.

7.
Answers will vary.

9.
Pearson's Product Moment Correlation Coefficient.

11.
There are many other possibilities, such as chance, relationship to a third variable, etc.

13.

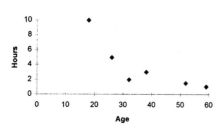

$\sum x = 225$
$\sum y = 22.5$
$\sum x^2 = 9653$
$\sum y^2 = 141.25$
$\sum xy = 625$
$n = 6$

$$r = \frac{n(\sum xy)-(\sum x)(\sum y)}{\sqrt{[n(\sum x^2)-(\sum x)^2]\,[n(\sum y^2)-(\sum y)^2]}}$$

$$r = \frac{6(625)-(225)(22.5)}{\sqrt{[6(9653)-(225)^2]\,[6(141.25)-(22.5)^2]}}$$

$r = -0.832$

13. continued
$H_0: \rho = 0$ and $H_1: \rho \neq 0$; C. V. $= \pm 0.811$; d. f. $= 4$; Decision: reject; there is a significant relationship between a person's age and the number of hours a person watches television.

15.

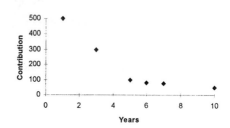

$\sum x = 32$
$\sum y = 1105$
$\sum x^2 = 220$
$\sum y^2 = 364,525$
$\sum xy = 3405$
$n = 6$

$$r = \frac{n(\sum xy)-(\sum x)(\sum y)}{\sqrt{[n(\sum x^2)-(\sum x)^2]\,[n(\sum y^2)-(\sum y)^2]}}$$

$$r = \frac{6(3405)-(32)(1105)}{\sqrt{[6(220)-(32)^2][6(364525)-(1105)^2]}}$$

$r = -0.883$

$H_0: \rho = 0$
$H_1: \rho \neq 0$
C. V. $= \pm 0.811$ d. f. $= 4$
Decision: Reject. There is a significant relationship between a person's age and his or her contribution.

17.

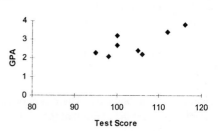

$\sum x = 832$
$\sum y = 22.1$
$\sum x^2 = 86,890$
$\sum y^2 = 63.83$
$\sum xy = 2321.1$
$n = 8$

$$r = \frac{n(\sum xy)-(\sum x)(\sum y)}{\sqrt{[n(\sum x^2)-(\sum x)^2]\,[n(\sum y^2)-(\sum y)^2]}}$$

17. continued

$$r = \frac{8(2321.1)-(832)(22.1)}{\sqrt{[8(86,890)-(832)^2][8(63.83)-(22.1)^2]}}$$

$r = 0.716$

H_0: $\rho = 0$
H_1: $\rho \neq 0$
C. V. = ± 0.707 d. f. = 6

Decision: Reject.   There is a significant relationship between test scores and G.P. A.

19.

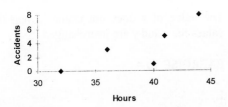

$\sum x = 193$
$\sum y = 17$
$\sum x^2 = 7537$
$\sum y^2 = 99$
$\sum xy = 705$
n = 5

$$r = \frac{n(\sum xy)-(\sum x)(\sum y)}{\sqrt{[n(\sum x^2)-(\sum x)^2][n(\sum y^2)-(\sum y)^2]}}$$

$r = 0.814$

H_0: $\rho = 0$
H_1: $\rho \neq 0$
C. V. = ± 0.878 d. f. = 3

Decision: Do not reject. There is not a significant relationship between the variables.

21.

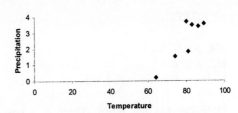

$\sum x = 1862$
$\sum y = 3222$
$\sum x^2 = 1,026,026$
$\sum y^2 = 3,009,596$

21. continued
$\sum xy = 1,754,975$
n = 6

$$r = \frac{n(\sum xy)-(\sum x)(\sum y)}{\sqrt{[n(\sum x^2)-(\sum x)^2][n(\sum y^2)-(\sum y)^2]}}$$

$$r = \frac{6(1,754,975) - (1862)(3222)}{\sqrt{[6(1,026,026) - 1862^2][6(3,009,596) - 3222^2]}}$$

$r = 0.997$

H_0: $\rho = 0$
H_1: $\rho \neq 0$
C. V. = ± 0.811 d. f. = 4

Decision: Reject. There is a significant linear relationship between the under 5 age group and the 65 and over age group.

23.

$\sum x = 557$
$\sum y = 17.7$
$\sum x^2 = 44,739$
$\sum y^2 = 55.99$
$\sum xy = 1468.9$
n = 7

$$r = \frac{n(\sum xy)-(\sum x)(\sum y)}{\sqrt{[n(\sum x^2)-(\sum x)^2][n(\sum y^2)-(\sum y)^2]}}$$

$$r = \frac{7(1468.9) - (557)(17.7)}{\sqrt{[7(44,739) - 557^2][7(55.99) - 17.7^2]}}$$

$r = 0.883$

H_0: $\rho = 0$
H_1: $\rho \neq 0$
C. V. = ± 0.754 d. f. = 5

Decision: Reject. There is a significant linear relationship between temperature and precipitation.

25.

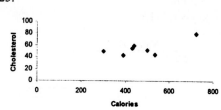

$\sum x = 3315$
$\sum y = 385$
$\sum x^2 = 1,675,225$
$\sum y^2 = 22,103$
$\sum xy = 189,495$
$n = 7$

$$r = \frac{n(\sum xy)-(\sum x)(\sum y)}{\sqrt{[n(\sum x^2)-(\sum x)^2]\,[n(\sum y^2)-(\sum y)^2]}}$$

$$r = \frac{7(189,495)-(3315)(385)}{\sqrt{[7(1,675,225)-(3315)^2][7(22,103)-(385)^2]}}$$

$r = 0.725$

$H_0: \rho = 0$
$H_1: \rho \neq 0$
C. V. $= \pm 0.754$ d. f. $= 5$

Decision: Do not reject. There is a not a significant linear relationship between calories and cholesterol.

27.

$\sum x = 31$
$\sum y = 29$
$\sum x^2 = 187$
$\sum y^2 = 183$
$\sum xy = 119$
$n = 6$

$$r = \frac{n(\sum xy)-(\sum x)(\sum y)}{\sqrt{[n(\sum x^2)-(\sum x)^2]\,[n(\sum y^2)-(\sum y)^2]}}$$

$r = -0.909$

$H_0: \rho = 0$
$H_1: \rho \neq 0$
C. V. $= \pm 0.811$ d. f. $= 4$

27. continued
Decision: Reject.
There is a significant relationship between the years of service and the number of resignations.

29.

$$r = \frac{n(\sum xy)-(\sum x)(\sum y)}{\sqrt{[n(\sum x^2)-(\sum x)^2]\,[n(\sum y^2)-(\sum y)^2]}}$$

$$r = \frac{5(125)-(15)(35)}{\sqrt{[5(55)-(15)^2][5(285)-(35)^2]}} = 1$$

$$r = \frac{5(125)-(35)(15)}{\sqrt{[5(285)-(35)^2][5(55)-(15)^2]}} = 1$$

The value of r does not change when the values for x and y are interchanged.

EXERCISE 10-4

1.
Draw the scatter plot and test the significance of the correlation coefficient.

3.
$y' = a + bx$

5.
It is the line that is drawn through the points on the scatter plot such that the sum of the squares of the vertical distances each point is from the line is at a minimum.

7.
When r is positive, b will be positive. When r is negative, b will be negative.

9.
The closer r is to $+1$ or -1, the more accurate the predicted value will be.

11.
When r is not significant, the mean of the y' values should be used to predict y.

13.
$$a = \frac{(\sum y)(\sum x^2)-(\sum x)(\sum xy)}{n(\sum x^2)-(\sum x)^2}$$

$$a = \frac{(22.5)(9653)-(225)(625)}{6(9653)-(225)^2} = 10.499$$

$$b = \frac{n(\sum xy)-(\sum x)(\sum y)}{n(\sum x^2)-(\sum x)^2}$$

$$b = \frac{6(625)-(225)(22.5)}{6(9653)-(225)^2} = -0.18$$

13. continued

$y' = a + bx$

$y' = 10.499 - 0.18x$

$y' = 10.499 - 0.18(35) = 4.199$ hours

15.

$a = \frac{(\sum y)(\sum x^2) - (\sum x)(\sum xy)}{n(\sum x^2) - (\sum x)^2}$

$a = \frac{(1105)(220) - (32)(3405)}{6(220) - (32)^2}$

$a = \frac{243100 - 108960}{1320 - 1024} = \frac{134140}{296} = 453.176$

$b = \frac{n(\sum xy) - (\sum x)(\sum y)}{n(\sum x^2) - (\sum x)^2}$

$b = \frac{6(3405) - (32)(1105)}{6(220) - (32)^2} = \frac{20430 - 35360}{296}$

$b = \frac{-14930}{296} = -50.439$

$y' = a + bx$

$y' = 453.176 - 50.439x$

$y' = 453.176 - 50.439(4) = \$251.42$

17.

$a = \frac{(\sum y)(\sum x^2) - (\sum x)(\sum xy)}{n(\sum x^2) - (\sum x)^2}$

$a = \frac{(22.1)(86890) - (832)(2321.1)}{8(86890) - (832)^2}$

$a = \frac{1920269 - 1931155.2}{695120 - 692224} = \frac{10886.2}{2896} = -3.759$

$b = \frac{n(\sum xy) - (\sum x)(\sum y)}{n(\sum x^2) - (\sum x)^2}$

$b = \frac{8(2321.1) - (832)(22.1)}{8(86890) - (832)^2}$

$b = \frac{18568.8 - 18387.2}{2986} = \frac{181.6}{2896} = 0.063$

$y' = a + bx$

$y' = -3.759 + 0.063x$

$y' = -3.759 + 0.063(104) = 2.793$

19.

Since r is not significant, the regression line should not be computed.

21.

$a = \frac{(\sum y)(\sum x^2) - (\sum x)(\sum xy)}{n(\sum x^2) - (\sum x)^2}$

$a = \frac{(3222)(1026026) - (1862)(1754975)}{6(1026026) - (1862)^2}$

$a = 14.165$

21. continued

$b = \frac{n(\sum xy) - (\sum x)(\sum y)}{n(\sum x^2) - (\sum x)^2}$

$b = \frac{6(1754975) - (1862)(3222)}{6(1026026) - (1862)^2} = 1.685$

$y' = a + bx$

$y' = 14.165 + 1.685x$

$y' = 14.165 + 1.685(200) = 351$ under 5.

23.

$a = \frac{(\sum y)(\sum x^2) - (\sum x)(\sum xy)}{n(\sum x^2) - (\sum x)^2}$

$a = \frac{(17.7)(44739) - (557)(1468.9)}{7(44739) - (557)^2} = -8.994$

$b = \frac{n(\sum xy) - (\sum x)(\sum y)}{n(\sum x^2) - (\sum x)^2}$

$b = \frac{7(1468.9) - (557)(17.7)}{7(44739) - (557)^2} = 0.1448$

$y' = a + bx$

$y' = -8.994 + 0.1448x$

$y' = -8.994 + 0.1448(70) = 1.1$ inches

25.

Since r is not significant, no regression should be done.

27.

$a = \frac{(\sum y)(\sum x^2) - (\sum x)(\sum xy)}{n(\sum x^2) - (\sum x)^2}$

$a = \frac{(29)(187) - (31)(119)}{6(187) - (31)^2} = 10.770$

$b = \frac{n(\sum xy) - (\sum x)(\sum y)}{n(\sum x^2) - (\sum x)^2}$

$b = \frac{6(119) - (31)(29)}{6(187) - (31)^2} = -1.149$

$y' = a + bx$

$y' = 10.770 - 1.149x$

$y' = 10.770 - 1.149(4) = 6.174$

29.

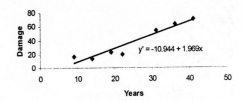

$\sum x = 172$

29. continued

$\sum y = 262$

$\sum x^2 = 5060$

$\sum y^2 = 13340$

$\sum xy = 8079$

$n = 7$

$r = \dfrac{n(\sum xy)-(\sum x)(\sum y)}{\sqrt{[n(\sum x^2)-(\sum x)^2][n(\sum y^2)-(\sum y)^2]}}$

$r = \dfrac{7(8079)-(172)(262)}{\sqrt{[7(5060)-(172)^2][7(13340)-(262)^2]}} = 0.956$

$H_0:\ \rho = 0$

$H_1:\ \rho \neq 0$

C. V. $= \pm 0.754$ d. f. $= 5$

Decision: Reject

There is a significant relationship between the number of years a person smokes and the amount of lung damage.

$a = \dfrac{(\sum y)(\sum x^2)-(\sum x)(\sum xy)}{n(\sum x^2)-(\sum x)^2}$

$a = \dfrac{(262)(5060)-(172)(8079)}{7(5060)-(172)^2} = -10.944$

$b = \dfrac{n(\sum xy)-(\sum x)(\sum y)}{n(\sum x^2)-(\sum x)^2}$

$b = \dfrac{7(8079)-(172)(262)}{7(5060)-(172)^2} = 1.969$

$y' = a + bx$

$y' = -10.944 + 1.969x$

$y' = -10.944 + 1.969(30) = 48.126$

31.

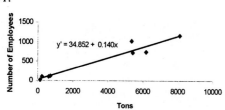

$\sum x = 26{,}728$

$\sum y = 4027$

$\sum x^2 = 162{,}101{,}162$

$\sum y^2 = 3{,}550{,}103$

$\sum xy = 23{,}663{,}669$

$n = 8$

$r = \dfrac{n(\sum xy)-(\sum x)(\sum y)}{\sqrt{[n(\sum x^2)-(\sum x)^2][n(\sum y^2)-(\sum y)^2]}}$

$r = \dfrac{8(23662669)-(26728)(4027)}{\sqrt{[8(162101162)-26728^2][8(3550103)-(4027)^2]}}$

31. continued

$r = 0.970$

$H_0:\ \rho = 0$

$H_1:\ \rho \neq 0$

C. V. $= \pm 0.707$ d. f. $= 6$

Decision: Reject.

There is a significant relationship between the number of tons of coal produced and the number of employees.

$a = \dfrac{(\sum y)(\sum x^2)-(\sum x)(\sum xy)}{n(\sum x^2)-(\sum x)^2}$

$a = \dfrac{(4027)(162101162)-(26728)(23663669)}{8(162101162)-(26728)^2}$

$a = 34.852$

$b = \dfrac{n(\sum xy)-(\sum x)(\sum y)}{n(\sum x^2)-(\sum x)^2}$

$b = \dfrac{8(23663669)-(26728)(4027)}{8(162101162)-(26728)^2} = 0.140$

$y' = a + bx$

$y' = 34.852 + 0.140x$

$y' = 34.852 + 0.140(500) = 104.8$

33.

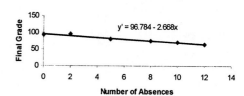

$\sum x = 37$

$\sum y = 482$

$\sum x^2 = 337$

$\sum y^2 = 39526$

$\sum xy = 2682$

$n = 6$

$r = \dfrac{n(\sum xy)-(\sum x)(\sum y)}{\sqrt{[n(\sum x^2)-(\sum x)^2][n(\sum y^2)-(\sum y)^2]}}$

$r = \dfrac{6(2682)-(37)(482)}{\sqrt{[6(337)-(37)^2][6(39526)-(482)^2]}}$

$r = -0.981$

$H_0:\ \rho = 0$

$H_1:\ \rho \neq 0$

C. V. $= \pm 0.811$ d. f. $= 4$

33. continued

Decision: Reject.

There is a significant negative relationship between the number of absences and the final grade.

$$a = \frac{(\sum y)(\sum x^2) - (\sum x)(\sum xy)}{n(\sum x^2) - (\sum x)^2}$$

$$a = \frac{(482)(337) - (37)(2682)}{6(337) - (37)^2} = 96.784$$

$$b = \frac{n(\sum xy) - (\sum x)(\sum y)}{n(\sum x^2) - (\sum x)^2}$$

$$b = \frac{6(2682) - (37)(482)}{6(337) - (37)^2} = -2.668$$

$y' = a + bx$

$y' = 96.784 - 2.668x$

35.

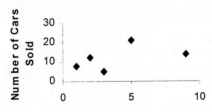

Years of Experience

$\sum x = 20$
$\sum y = 60$
$\sum x^2 = 120$
$\sum y^2 = 870$
$\sum xy = 278$
$n = 5$

$$r = \frac{n(\sum xy) - (\sum x)(\sum y)}{\sqrt{[n(\sum x^2) - (\sum x)^2][n(\sum y^2) - (\sum y)^2]}}$$

$$r = \frac{5(278) - (20)(60)}{\sqrt{[5(120) - 20^2][5(870) - 60^2]}}$$

$r = 0.491$

$H_0: \rho = 0$
$H_1: \rho \neq 0$
t = 0.976; 0.20 < P-value < 0.50 (0.401)

Decision: Do not reject since P-value > 0.05. There is no significant relationship between the number of years of experience and the number of cars sold per month. Since r is not significant, no regression analysis should be done.

37.

For 11 – 15

$\overline{x} = 5.3333$
$\overline{y} = 184.1667$
$b = -50.439$
$a = \overline{y} - b\overline{x}$
$a = 184.1667 - (-50.439)(5.3333)$
$a = 184.1667 + 269.0063$
$a = 453.173$ (differs due to rounding)

For 11 – 16

$\overline{x} = 40.33$
$\overline{y} = 8.33$
$b = -0.317$
$a = \overline{y} - b\overline{x}$
$a = 8.33 - (-0.317)(40.33)$
$a = 8.33 + 12.78$
$a = 21.11$ or 21.1

For 11 – 17

$\overline{x} = 104$
$\overline{y} = 2.7625$
$b = 0.063$
$a = \overline{y} - b\overline{x}$
$a = 2.7625 - (0.063)(104)$
$a = 2.7625 - 6.552$
$a = -3.7895$ (differs due to rounding)

EXERCISE SET 10-5

1.
Explained variation is the variation obtained from the predicted y' values, and is computed by $\sum(y' - \overline{y})^2$.

3.
Total variation is the sum of the explained and unexplained variation and is computed by $\sum(y - \overline{y})^2 = \sum(y' - \overline{y})^2 + \sum(y - y')^2$.

5.
It is found by squaring r.

7.
The coefficient of non-determination is $1 - r^2$.

9.
For $r = 0.70$, $r^2 = 0.49$, $1 - r^2 = 0.51$
49% of the variation of y is due to the variation of x, and 51% of the variation of y is due to chance.

11.
For $r = 0.37$, $r^2 = 0.1369$, $1 - r^2 = 0.8631$
13.69% of the variation of y is due to the variation of x, and 86.31% of the variation of y is due to chance.

13.
For $r = 0.05$, $r^2 = 0.0025$, $1 - r^2 = 0.9975$
0.25% of the variation of y is due to the variation of x, and 99.75% of the variation of y is due to chance.

15.
$$S_{est} = \sqrt{\frac{\sum y^2 - a\sum y - b\sum xy}{n-2}}$$

$$S_{est} = \sqrt{\frac{141.25 - 10.499(22.5) - (-0.18)(625)}{6-2}}$$

$$S_{est} = \sqrt{4.380625}$$

$$s_{est} = 2.09$$

17.
$$S_{est} = \sqrt{\frac{\sum y^2 - a\sum y - b\sum xy}{n-2}} =$$

$$S_{est} = \sqrt{\frac{364525 - (453.176)(1105) - (-50.439)(3405)}{6-2}}$$

$$S_{est} = 94.22$$

19.
$y' = 10.499 - 0.18x$
$y' = 10.499 - 0.18(20)$
$y' = 6.899$

$$y' - t_{\frac{\alpha}{2}} \cdot S_{est}\sqrt{1 + \frac{1}{n} + \frac{n(x-\overline{X})}{n\sum x^2 - (\sum x)^2}} < y < y' +$$

$$t_{\frac{\alpha}{2}} \cdot S_{est}\sqrt{1 + \frac{1}{n} + \frac{n(x-\overline{X})^2}{n\sum x^2 - (\sum x)^2}}$$

$$6.899 - (2.132)(2.09)\sqrt{1 + \frac{1}{6} + \frac{6(20-37.5)^2}{6(9653)-225^2}}$$

$$< y < 6.899 +$$
$$(2.132)(2.09)\sqrt{1 + \frac{1}{6} + \frac{6(20-37.5)^2}{6(9653)-225^2}}$$

$6.899 - (2.132)(2.09)(1.191) < y < 6.899 +$
$\quad\quad (2.132)(2.09)(1.91)$
$1.59 < y < 12.21$

21.
$y' = 453.176 - 50.439x$
$y' = 453.176 - 50.439(4)$

21. continued
$y' = 251.42$

$$y' - t_{\frac{\alpha}{2}} \cdot S_{est}\sqrt{1 + \frac{1}{n} + \frac{n(x-\overline{X})^2}{n\sum x^2 - (\sum x)^2}} < y <$$

$$y' + t_{\frac{\alpha}{2}} \cdot S_{est}\sqrt{1 + \frac{1}{n} + \frac{n(x-\overline{X})^2}{n\sum x^2 - (\sum x)^2}}$$

$$251.42 - 2.132(94.22)\sqrt{1 + \frac{1}{6} + \frac{6(4-5.33)^2}{6(220)-32^2}}$$

$$< y < 251.42 + 2.132(94.22)\sqrt{1 + \frac{1}{6} + \frac{6(4-5.33)^2}{6(220)-32^2}}$$

$251.42 - (2.132)(94.22)(1.1) < y <$
$\quad\quad 251.42 + (2.132)(94.22)(1.1)$
$\$30.46 < y < \$472.38$

EXERCISE SET 10-6

1.
Simple linear regression has one independent variable and one dependent variable. Multiple regression has one dependent variable and two or more independent variables.

3.
The relationship would include all variables.

5.
The multiple correlation coefficient R is always higher than the individual correlation coefficients. Also, the value of R can range from 0 to $+1$.

7.
$y' = 9.6 + 2.2x_1 - 1.08x_2$
$y' = 9.6 + 2.2(9) - 1.08(24) = 3.48$

9.
$y' = 5000 + 97x_1 + 35x_2$
$y' = 5000 + 97(120) + 35(650)$
$y' = \$39,390$

11.
R is a measure of the strength of the relationship between the dependent variables and all the independent variables.

13.
R^2 is the coefficient of multiple determination. R^2_{adj} is adjusted for sample size.

15.
The F test is used to test the significance of R.

REVIEW EXERCISES - CHAPTER 10

1.

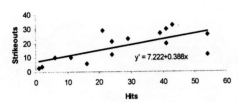

$\sum x = 406$
$\sum y = 266$
$\sum x^2 = 15{,}416$
$\sum y^2 = 6154$
$\sum xy = 8919$
n = 15

$r = \dfrac{n(\sum xy) - (\sum x)(\sum y)}{\sqrt{[n(\sum x^2) - (\sum x)^2]\,[n(\sum y^2) - (\sum y)^2]}}$

$r = \dfrac{15(8919) - (406)(266)}{\sqrt{[15(15416) - (406)^2][15(6154) - (266)^2]}}$

$r = 0.682$

$H_0\colon \rho = 0$
$H_1\colon \rho \neq 0$
C. V. = ± 0.641 d. f. = 13

Decision: Reject. There is a significant relationship between hits and strikeouts.

$a = \dfrac{(\sum y)(\sum x^2) - (\sum x)(\sum xy)}{n(\sum x^2) - (\sum x)^2}$

$a = \dfrac{(266)(15416) - (406)(8919)}{15(15416) - (406)^2} = 7.222$

$b = \dfrac{n(\sum xy) - (\sum x)(\sum y)}{n(\sum x^2) - (\sum x)^2}$

$b = \dfrac{15(8919) - (406)(266)}{15(15416) - (406)^2} = 0.388$

$y' = a + bx$
$y' = 7.222 + 0.388x$
$y' = 7.222 + 0.388(30) = 18.86$ or 19 strikeouts

3.

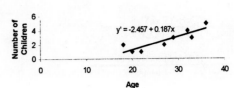

$\sum x = 217$
$\sum y = 21$
$\sum x^2 = 6187$
$\sum y^2 = 69$
$\sum xy = 626$
n = 8

$r = \dfrac{n(\sum xy) - (\sum x)(\sum y)}{\sqrt{[n(\sum x^2) - (\sum x)^2]\,[n(\sum y^2) - (\sum y)^2]}}$

$r = \dfrac{8(626) - (217)(21)}{\sqrt{[8(6187) - (217)^2][8(69) - (21)^2]}}$

$r = 0.873$

$H_0\colon \rho = 0$
$H_1\colon \rho \neq 0$
C. V. = ± 0.834 d. f. = 6
Decision: Reject. There is a significant relationship between the mother's age and the number of children she has.

$a = \dfrac{(\sum y)(\sum x^2) - (\sum x)(\sum xy)}{n(\sum x^2) - (\sum x)^2}$

$a = \dfrac{(21)(6187) - (217)(626)}{8(6187) - (217)^2} = -2.457$

$b = \dfrac{n(\sum xy) - (\sum x)(\sum y)}{n(\sum x^2) - (\sum x)^2}$

$b = \dfrac{8(626) - (217)(21)}{8(6187) - (217)^2} = 0.187$

$y' = a + bx$
$y' = -2.457 + 0.187x$
$y' = -2.457 + 0.187(34) = 3.9$

5.

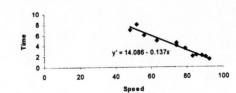

$\sum x = 884$
$\sum y = 47.8$
$\sum x^2 = 67728$
$\sum y^2 = 242.06$
$\sum xy = 3163.8$

5. continued
n = 12

$$r = \frac{n(\sum xy)-(\sum x)(\sum y)}{\sqrt{[n(\sum x^2)-(\sum x)^2]\,[n(\sum y^2)-(\sum y)^2]}}$$

$$r = \frac{12(3163.8)-(884)(47.8)}{\sqrt{[12(67728)-(884)^2][12(242.06)-(47.8)^2]}}$$

$$r = -0.974$$

H_0: $\rho = 0$
H_1: $\rho \neq 0$
C. V. = ± 0.708 d. f. = 10

Decision: Reject the null. There is a significant relationship between speed and time.

$$a = \frac{(\sum y)(\sum x^2)-(\sum x)(\sum xy)}{n(\sum x^2)-(\sum x)^2}$$

$$a = \frac{(47.8)(67728)-(884)(3163.8)}{12(67728)-(884)^2}$$

$$a = 14.086$$

$$b = \frac{n(\sum xy)-(\sum x)(\sum y)}{n(\sum x^2)-(\sum x)^2}$$

$$b = \frac{12(3163.8)-(884)(47.8)}{12(67728)-(884)^2}$$

$$b = -0.137$$

$y' = a + bx$
$y' = 14.086 - 0.137x$
$y' = 14.086 - 0.137(72) = 4.222$

7.

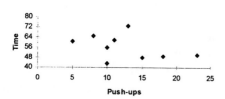

$\sum x = 113$
$\sum y = 507$
$\sum x^2 = 1657$
$\sum y^2 = 29309$
$\sum xy = 6198$
n = 9

$$r = \frac{n(\sum xy)-(\sum x)(\sum y)}{\sqrt{[n(\sum x^2)-(\sum x)^2]\,[n(\sum y^2)-(\sum y)^2]}}$$

$$r = \frac{9(6198)-(113)(507)}{\sqrt{[9(1657)-(113)^2][9(29309)-(507)^2]}}$$

$$r = -0.397$$

7. continued
H_0: $\rho = 0$
H_1: $\rho \neq 0$
C. V. = ± 0.798 d. f. = 7

Decision: Do not reject. Since the null hypothesis is not rejected, no regression should be done.

9.
$$S_{est} = \sqrt{\frac{\sum y^2 - a\sum y - b\sum xy}{n-2}}$$

$$S_{est} = \sqrt{\frac{242.06 - 14.086(47.8) + 0.137(3163.8)}{12-2}}$$

$$S_{est} = \sqrt{\frac{2.1898}{10}} = \sqrt{0.21898} = 0.468$$

11.
$y' = 14.086 - 0.137x$
$y' = 14.086 - 0.137(72) = 4.222$

$$y' - t_{\frac{\alpha}{2}} \cdot S_{est}\sqrt{1+\frac{1}{n}+\frac{n(x-\overline{X})^2}{n\Sigma x^2-(\Sigma x)^2}} < y <$$

$$y' + t_{\frac{\alpha}{2}} \cdot S_{est}\sqrt{1+\frac{1}{n}+\frac{n(x-\overline{X})^2}{n\Sigma x^2-(\Sigma x)^2}}$$

$$4.222 - 1.812(0.468)\sqrt{1+\frac{1}{12}+\frac{12(72-73.667)^2}{12(67,728)-884^2}}$$
$$< y < 4.222 + 1.812(0.468)\sqrt{1+\frac{1}{12}+\frac{12(72-73.667)^2}{12(67,728)-884^2}}$$

$$4.222 - 1.812(0.468)(1.041) < y <$$
$$4.222 + 1.812(0.468)(1.041)$$
$$3.34 < y < 5.10$$

13.
$y\prime = 12.8 + 2.09x_1 + 0.423x_2$
$y\prime = 12.8 + 2.09(4) + 0.423(2) = 22.006$

15.
$$R_{adj}^2 = 1 - \left[\frac{(1-R^2)(n-1)}{n-k-1}\right]$$

$$R_{adj}^2 = 1 - \left[\frac{(1-0.873^2)(10-1)}{10-3-1}\right]$$

$$R_{adj}^2 = 1 - \left[\frac{2.1408}{6}\right] = 0.643$$

CHAPTER 10 QUIZ

1. False, the y variable would decrease.
2. True
3. True

4. False, the relationship may be affected by another variable, or by chance.

5. False, a relationship may be caused by chance.

6. False, there are several independent variables and one dependent variable.

7. a.

8. a.

9. d.

10. c.

11. b.

12. scatter diagram

13. independent

14. $-1, +1$

15. b.

16. line of best fit

17. $+1, -1$

18.

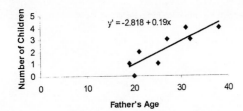

$\sum x = 213$
$\sum x^2 = 5985$
$\sum y = 18$
$\sum y^2 = 56$
$\sum xy = 539$
$n = 8$
$r = 0.857$
H_0: $\rho = 0$ and H_1: $\rho \neq 0$. C.V. $= \pm 0.707$ and d. f. $= 6$. Reject. There is a significant relationship between father's age and number of children.
$a = -2.818$ $b = 0.19$
$y' = -2.818 + 0.19x$
For $x = 35$ years old:
$y' = -2.818 + 0.19(35) = 3.8$ or 4

19.

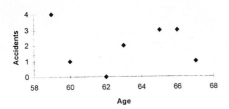

19. continued
$\sum x = 442$
$\sum x^2 = 27,964$
$\sum y = 14$
$\sum y^2 = 40$
$\sum xy = 882$
$n = 7$
$r = -0.078$
H_0: $\rho = 0$ and H_1: $\rho \neq 0$. d. f. $= 5$ and C. V. $= \pm 0.764$. Decision: do not reject. There is not a significant relationship between age and number of accidents.

20.

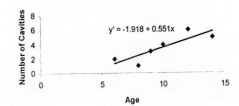

$\sum x = 59$
$\sum x^2 = 621$
$\sum y = 21$
$\sum y^2 = 91$
$\sum xy = 229$
$n = 6$
$r = 0.842$
H_0: $\rho = 0$ and H_1: $\rho \neq 0$; d. f. $= 4$ and C. V. $= \pm 0.811$. Decision: Reject. There is a significant relationship between age and number of cavities.
$a = -1.918$ $b = 0.551$
$y' = -1.918 + 0.551x$
When $x = 11$: $y' = -1.918 + 0.551(11)$
$y' = 4.14 \approx 4$ cavities

21.

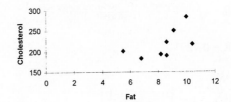

$\sum x = 67.2$
$\sum x^2 = 582.62$
$\sum y = 1740$
$\sum y^2 = 386,636$
$\sum xy = 14847.9$
$n = 8$
$r = 0.602$

21. continued

H_0: $\rho = 0$ and H_1: $\rho \neq 0$; d. f. = 6 and C.
V. $= \pm 0.707$. Decision: Do not reject.
There is no significant relationship between
fat and cholesterol.

22.

$$S_{est} = \sqrt{\frac{91-(-1.918)(21)-0.551(229)}{6-2}}$$
$$S_{est} = 1.129*$$

23.

(For calculation purposes only, since no
regression should be done.)

$$S_{est} = \sqrt{\frac{386,636-110.12(1740)-12.784(14,847.9)}{8-2}}$$
$$S_{est} = 29.47*$$

24.

$$y' = -1.918 + 0.551(7) = 1.939 \text{ or } 2$$

$$2 - 2.132(1.129)\sqrt{1 + \tfrac{1}{6} + \tfrac{6(11-9.833)^2}{6(621)-59^2}} < y$$

$$< 2 + 2.132(1.129)\sqrt{1 + \tfrac{1}{6} + \tfrac{6(11-9.833)^2}{6(621)-59^2}}$$

$$2 - 2.132(1.129)(1.095) < y < 2 + 2.132(1.129)(1.095)$$

$$-0.6 < y < 4.6 \Rightarrow 0 < y < 5*$$

25.

Since no regression should be done, the
prediction interval is $\bar{y} = 217.5$.

26.

$$y' = 98.7 + 3.82(3) + 6.51(1.5) = 119.9*$$

27.

$$R = \sqrt{\frac{(0.561)^2+(0.714)^2-2(0.561)(0.714)(0.625)}{1-(0.625)^2}}$$

$$R = 0.729*$$

28.

$$R_{adj}^2 = 1 - \left[\frac{(1-0.774^2)(8-1)}{(8-2-1)}\right]$$

$$R_{adj}^2 = 0.439*$$

*These answers may vary due to method of
calculation or rounding.

Note: Graphs are not to scale and are intended to convey a general idea.

Answers may vary due to rounding, TI-83's, or computer programs.

EXERCISE SET 11-2

1.

The variance test compares a sample variance to a hypothesized population variance, while the goodness of fit test compares a distribution obtained from a sample with a hypothesized distribution.

3.

The expected values are computed based on what the null hypothesis states about the distribution.

5.

H_0: The number of accidents is equally distributed throughout the week. (claim)
H_1: The distribution is not the same as stated in the null hypothesis.
C. V. = 12.592   d. f. = 6 $\alpha = 0.05$
$E = \frac{189}{7} = 27$

$\chi^2 = \sum \frac{(O-E)^2}{E} = \frac{(28-27)^2}{27} + \frac{(32-27)^2}{27} +$

$\frac{(15-27)^2}{27} + \frac{(14-27)^2}{27} + \frac{(38-27)^2}{27} + \frac{(43-27)^2}{27}$

$+ \frac{(19-27)^2}{27} = 28.887$

Alternate Solution:

O	E	O − E	(O − E)²	$\frac{(O-E)^2}{E}$
28	27	1	1	0.037
32	27	5	25	0.926
15	27	-12	144	5.333
14	27	-13	169	6.259
38	27	11	121	4.481
43	27	16	256	9.481
19	27	-8	64	2.370
				28.887

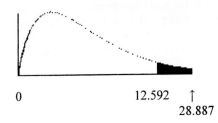

0   12.592   ↑
                              28.887

5. continued
Reject the null hypothesis. There is enough evidence to reject the claim that the number of accidents is equally distributed during the week.

7.
H_0: The proportions are distributed as follows: 28.1% purchased a small car, 47.8% purchased a mid-sized car, 7% purchased a large car, and 17.1% purchased a luxury car.
H_1: The distribution is not the same as stated in the null hypothesis. (claim)
C. V. = 7.815   d. f. = 3 $\alpha = 0.05$

$\chi^2 = \sum \frac{(O-E)^2}{E} = \frac{(25-28.1)^2}{28.1} + \frac{(50-47.8)^2}{47.8}$

$+ \frac{(10-7)^2}{7} + \frac{(15-17.1)^2}{17.1} = 1.9869$

Alternate Solution:

O	E	O − E	(O − E)²	$\frac{(O-E)^2}{E}$
25	28.1	-3.1	9.61	0.3420
50	47.8	2.2	4.84	0.1013
10	7	3	9	1.2857
15	17.1	-2.1	4.41	0.2579
				1.9869

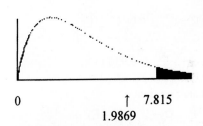

0   ↑ 7.815
                              1.9869

Do not reject the null hypothesis. There is not enough evidence to support the claim that the proportions are different.

9.
H_0: The proportions are distributed as follows: safe - 35%, not safe - 52%, no opinion - 13%.
H_1: The distribution is not the same as stated in the null hypothesis. (claim)
C. V. = 9.210   d. f. = 2 $\alpha = 0.01$

$\chi^2 = \frac{(40-42)^2}{42} + \frac{(60-62.4)^2}{62.4} + \frac{(20-15.6)^2}{15.6}$

$\chi^2 = 1.4285$

9. continued
Alternate Solution:

O	E	O − E	$(O − E)^2$	$\frac{(O-E)^2}{E}$
40	42	-2	4	0.0952
60	62.4	-2.4	5.76	0.0923
20	15.6	4.4	19.36	1.2410
				1.4285

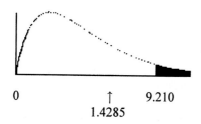

0   ↑   9.210
        1.4285

Do not reject the null hypothesis. There is not enough evidence to support the claim that the proportions are different.

11.
H_0: The distribution of loans is as follows: 21% - mortgages, 39% - autos, 20% - unsecured, 12% - real estate, and 8% - miscellaneous. (claim)
H_1: The distribution is not the same as stated in the null hypothesis.
C. V. = 9.488   d. f. = 4 $\alpha = 0.05$

$$\chi^2 = \frac{(25-21)^2}{21} + \frac{(44-39)^2}{39} + \frac{(18-20)^2}{20} +$$

$$\frac{(8-12)^2}{12} + \frac{(4-8)^2}{8} = 4.9362$$

Alternate Solution:

O	E	O − E	$(O − E)^2$	$\frac{(O-E)^2}{E}$
25	21	4	16	0.7619
44	39	5	25	0.6410
18	20	-2	4	0.2
8	12	-4	16	1.3333
4	8	-4	16	2.0000
				4.9362

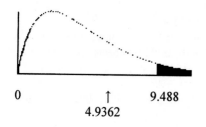

0   ↑   9.488
        4.9362

11. continued
Do not reject the null hypothesis. There is not enough evidence to support the claim that the distribution is different.

13.
H_0: The method of payment for purchases is distributed as follows: 53% cash, 30% checks, 16% credit cards, and 1% no preference. (claim)
H_1: The distribution is not the same as stated in the null hypothesis.
C. V. = 11.345   d. f. = 3 $\alpha = 0.01$

$$\chi^2 = \frac{(400-424)^2}{424} + \frac{(210-240)^2}{240} + \frac{(170-128)^2}{128}$$

$$+ \frac{(20-8)^2}{8} = 36.8898$$

Alternate Solution:

O	E	O − E	$(O − E)^2$	$\frac{(O-E)^2}{E}$
400	424	-24	576	1.3585
210	240	-30	900	3.7500
170	128	42	1764	13.7813
20	8	12	144	18.0000
				36.8898

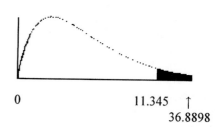

0   11.345 ↑
        36.8898

Reject the null hypothesis. There is enough evidence to reject the claim that the distribution is the same as reported in the survey.

15.
H_0: The distribution is as follows: violent offenses - 29.5%, property offenses - 29%, drug offenses - 30.2%, weapons offenses - 10.6%, other offenses - 0.7%. (claim)
H_1: The distribution is not the same as stated in the null hypothesis.
C. V. = 9.488   d. f. = 4 $\alpha = 0.05$

$$\chi^2 = \sum \frac{(O-E)^2}{E} = \frac{(298-295)^2}{295} + \frac{(275-290)^2}{290}$$

$$+ \frac{(344-302)^2}{302} + \frac{(80-106)^2}{106} + \frac{(3-7)^2}{7}$$

15. continued

$\chi^2 = 15.3106$

Alternate Solution:

O	E	O − E	$(O-E)^2$	$\frac{(O-E)^2}{E}$
298	295	3	9	0.0305
275	290	-15	225	0.7759
344	302	42	1764	5.8411
80	106	-26	676	6.3774
3	7	-4	16	2.2857
				15.3106

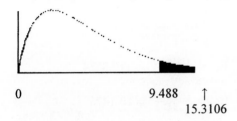

0   9.488   ↑
                              15.3106

Reject the null hypothesis. There is not enough evidence to support the claim that the proportions are as stated.

17.
H_0: 50% of customers purchase word processing programs, 25% purchase spread sheet programs, and 25% purchase data base programs. (claim)
H_1: The distribution is not the same as stated in the null hypothesis.
$\alpha = 0.05$ d. f. = 2
P-value > 0.10 (0.741)

$\chi^2 = \sum \frac{(O-E)^2}{E} = \frac{(38-40)^2}{40} + \frac{(23-20)^2}{20} +$
$\frac{(19-20)^2}{20} = 0.6$

Alternate Solution:

O	E	O − E	$(O-E)^2$	$\frac{(O-E)^2}{E}$
38	40	-2	4	0.1
23	20	3	9	0.45
19	20	-1	1	0.05
				0.60

Do not reject the null hypothesis since P-value > 0.05. There is not enough evidence to reject the store owner's assumption.

19.
Answers will vary.

EXERCISE SET 11-3

1.
The independence test and the goodness of fit test both use the same formula for computing the test-value; however, the independence test uses a contingency table whereas the goodness of fit test does not.

3.
H_0: The variables are independent or not related.
H_1: The variables are dependent or related.

5.
The expected values are computed as (row total · column total) ÷ grand total.

7.
H_0: $p_1 = p_2 = p_3 = \cdots = p_n$
H_1: At least one proportion is different from the others.

9.
H_0: Type of pet owned is independent of annual household income.
H_1: Type of pet owned is dependent on annual household income. (claim)

C. V. = 21.026   d. f. = 12 $\alpha = 0.05$

$E = \frac{(\text{row sum})(\text{column sum})}{\text{grand total}}$

$E_{1,1} = \frac{(534)(1003)}{4004} = 133.7667$

$E_{1,2} = \frac{(534)(1000)}{4004} = 133.3666$

$E_{1,3} = \frac{(534)(1000)}{4004} = 133.3666$

$E_{1,4} = \frac{(534)(1001)}{4004} = 133.5$

$E_{2,1} = \frac{(800)(1003)}{4004} = 200.3996$

$E_{2,2} = \frac{(800)(1000)}{4004} = 199.8002$

$E_{2,3} = \frac{(800)(1000)}{4004} = 199.8002$

$E_{2,4} = \frac{(800)(1001)}{4004} = 200.0$

$E_{3,1} = \frac{(869)(1003)}{4004} = 208.6661$

$E_{3,2} = \frac{(833)(1000)}{4004} = 208.0420$

9. continued

$$E_{3,3} = \frac{(833)(1000)}{4004} = 208.0420$$

$$E_{3,4} = \frac{(833)(1001)}{4004} = 208.25$$

$$E_{4,1} = \frac{(968)(1003)}{4004} = 242.2835$$

$$E_{4,2} = \frac{(968)(1000)}{4004} = 241.7582$$

$$E_{4,3} = \frac{(968)(1000)}{4004} = 241.7582$$

$$E_{4,4} = \frac{(968)(1001)}{4004} = 242.0$$

Type of Pet

Income	Dog	Cat
< $12,500	127(133.7667)	139(133.3666)
$12,500 - $24,999	191(199.8002)	197(199.8002)
$25,000 - $39,999	216(217.6841)	215(217.0330)
$40,000 - $59,999	215(208.6661)	212(208.0420)
$60,000 & over	254(242.4835)	237(241.7582)

Income	Bird	Horse
< $12,500	173(133.3666)	95(133.5)
$12,500 - $24,999	209(199.8002)	203(200.0)
$25,000 -$39,999	220(217.0330)	218(217.25)
$40,000 - $59,999	175(208.0420)	231(208.25)
$60,000 & over	223(241.7582)	254(242.0)

$$\chi^2 = \sum \frac{(O-E)^2}{E} = \frac{(127-133.7667)^2}{133.7667} + \frac{(139-133.3666)^2}{133.3666}$$

$$+ \frac{(173-133.3666)^2}{133.3666} + \frac{(95-133.5)^2}{133.5} + \frac{(191-199.8002)^2}{199.8002}$$

$$+ \frac{(197-199.8002)^2}{199.8002} + \frac{(209-199.8002)^2}{199.8002} + \frac{(203-200.0)^2}{200}$$

$$+ \frac{(216-217.6841)^2}{217.6841} + \frac{(215-217.0330)^2}{217.0330} + \frac{(220-217.0330)^2}{217.0330}$$

$$+ \frac{(218-217.25))^2}{217.25} + \frac{(215-208.6661)^2}{208.6661} + \frac{(212-208.0420)^2}{208.0420}$$

$$+ \frac{(175-208.0420)^2}{208.0420} + \frac{(231-208.25)^2}{208.25} + \frac{(254-242.4835)^2}{242.4835}$$

$$+ \frac{(237-241.7582)^2}{241.7582} + \frac{(223-241.7582)^2}{241.7582} + \frac{(254-242.0)^2}{242}$$

$$\chi^2 = 35.177$$

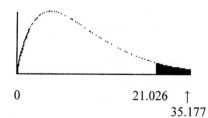

0   21.026 ↑
                35.177

9. continued
Reject the null hypothesis. There is enough evidence to support the claim that the type of pet is dependent upon the income of the owner.

11.
H_0: The composition of the House of Representatives is independent of the state.
H_1: The composition of the House of Representatives is dependent upon the state. (claim)

C. V. = 7.815   d. f. = 3 $\alpha = 0.05$

$$E = \frac{(\text{row sum})(\text{column sum})}{\text{grand total}}$$

$$E_{1,1} = \frac{(203)(320)}{542} = 119.8524$$

$$E_{1,2} = \frac{(203)(222)}{542} = 83.1476$$

$$E_{2,1} = \frac{(98)(320)}{542} = 57.8598$$

$$E_{2,2} = \frac{(98)(222)}{542} = 40.1402$$

$$E_{3,1} = \frac{(100)(320)}{542} = 59.0406$$

$$E_{3,2} = \frac{(100)(222)}{542} 40.9594$$

$$E_{4,1} = \frac{(141)(320)}{542} = 83.2472$$

$$E_{4,2} = \frac{(141)(222)}{542} = 57.7528$$

State	Democrats	Republicans
PA	100(119.8524)	103(83.1476)
OH	39(57.8598)	59(40.1402)
WV	75(59.0406)	25(40.9594)
MD	106(83.2472)	35(57.7528)

$$\chi^2 = \sum \frac{(O-E)^2}{E} = \frac{(100-119.8524)^2}{119.8524} + \frac{(103-83.1476)^2}{83.1476}$$

$$+ \frac{(39-57.8598)^2}{57.8598} + \frac{(59-40.1402)^2}{40.1402} + \frac{(75-59.0406)^2}{59.0406}$$

$$+ \frac{(25-40.9594)^2}{40.9594} + \frac{(106-83.2472)^2}{83.2472} + \frac{(35-57.7528)^2}{57.7528}$$

$$\chi^2 = 48.7521$$

11. continued

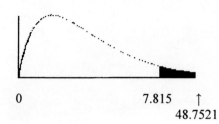

0   7.815   ↑
                                  48.7521

Reject the null hypothesis. There is enough evidence to support the claim that the composition is dependent upon the state.

13.
H_0: The number of ads people think they've seen or heard in the media is independent of the gender of the individual.
H_1: The number of ads people think they've seen or heard in the media is dependent upon the gender of the individual. (claim)
C. V. = 13.277   d. f. = 4 $\alpha = 0.01$

$E_{1,1} = \frac{(300)(95)}{510} = 55.882$

$E_{1,2} = \frac{(300)(110)}{510} = 64.706$

$E_{1,3} = \frac{(300)(144)}{510} = 84.706$

$E_{1,4} = \frac{(300)(84)}{510} = 49.412$

$E_{1,5} = \frac{(300)(77)}{510} = 45.294$

$E_{2,1} = \frac{(210)(95)}{510} = 39.118$

$E_{2,2} = \frac{(210)(110)}{510} = 45.294$

$E_{2,3} = \frac{(210)(144)}{510} = 59.294$

$E_{2,4} = \frac{(210)(84)}{510} = 34.588$

$E_{2,5} = \frac{(210)(77)}{510} = 31.706$

Gender	1 - 30	31 - 50	51 - 100
Men	45(55.882)	60(64.706)	90(84.706)
Women	50(39.118)	50(45.294)	54(59.294)
Total	95	110	144

Gender	101 - 300	301 or more	Total
Men	54(49.412)	51(45.294)	300
Women	30(34.588)	26(31.706)	210
Total	84	77	510

13. continued

$$\chi^2 = \sum \frac{(O-E)^2}{E} = \frac{(45\text{-}55.882)^2}{55.882} + \frac{(60\text{-}64.706)^2}{64.706}$$

$$+ \frac{(90\text{-}84.706)^2}{84.706} + \frac{(54\text{-}49.412)^2}{49.412} + \frac{(51\text{-}45.294)^2}{45.294}$$

$$+ \frac{(50\text{-}39.118)^2}{39.118} + \frac{(50\text{-}45.294)^2}{45.294} + \frac{(54\text{-}59.294)^2}{59.294}$$

$$+ \frac{(30\text{-}34.588)^2}{34.588} + \frac{(26\text{-}31.706)^2}{31.706} = 9.562$$

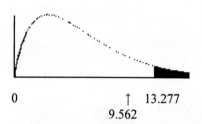

0   ↑   13.277
                   9.562

Do not reject the null hypothesis. There is not enough evidence to support the claim that the number of ads people think they've seen or heard is related to the gender of the individual.

15.
H_0: The grade a student receives is independent of the number of hours the student works.
H_1: The grade a student receives is dependent upon the number of hours the student works. (claim)

C. V. = 12.592   d. f. = 6 $\alpha = 0.05$

$E_{1,1} = \frac{(21)(18)}{92} = 4.1087$

$E_{1,2} = \frac{(21)(40)}{92} = 9.1304$

$E_{1,3} = \frac{(21)(15)}{92} = 3.4239$

$E_{1,4} = \frac{(21)(19)}{92} = 4.3370$

$E_{2,1} = \frac{(27)(18)}{92} = 5.2826$

$E_{2,2} = \frac{(27)(40)}{92} 11.7391$

$E_{2,3} = \frac{(27)(15)}{92} = 4.4022$

$E_{2,4} = \frac{(27)(19)}{92} = 5.5761$

$E_{3,1} = \frac{(44)(18)}{92} = 8.6087$

15. continued

$$E_{3,2} = \frac{(44)(40)}{92} = 19.1304$$

$$E_{3,3} = \frac{(44)(15)}{92} = 7.1739$$

$$E_{3,4} = \frac{(44)(19)}{92} = 9.0870$$

Hours Working	A	B
20 +	5(4.1087)	8(9.1304)
10 - 20	5(5.2826)	12(11.7391)
< 10	8(8.6087)	20(19.1304)

Hours Working	C	D/F
20 +	3(3.4239)	5(4.3370)
10 - 20	6(4.4022)	4(5.5761)
< 10	6(7.1739)	10(9.0870)

$$\chi^2 = \sum \frac{(O-E)^2}{E} = \frac{(5-4.1087)^2}{4.1087} + \frac{(8-9.1304)^2}{9.1304}$$

$$+ \frac{(3-3.4239)^2}{3.4239} + \frac{(5-4.3370)^2}{4.3370} + \frac{(5-5.2826)^2}{5.2826}$$

$$+ \frac{(12-11.7391)^2}{11.7391} + \frac{(6-4.4022)^2}{4.4022} + \frac{(4-5.5761)^2}{5.5761}$$

$$+ \frac{(8-8.6087)^2}{8.6087} + \frac{(20-19.1304)^2}{19.1304} + \frac{(6-7.1739)^2}{7.1739}$$

$$+ \frac{(10-9.0870)^2}{9.0870} = 1.9$$

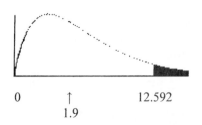

Do not reject the null hypothesis. There is not enough evidence to support the claim that grades are dependent upon the number of hours a student works.

17.
H_0: The type of video rented by a person is independent of the person's age.
H_1: The type of video a person rents is dependent on the person's age. (claim)
C. V. = 13.362   d. f. = 8 $\alpha = 0.10$

Age	Doc.	Comedy	Mystery
12-20	14(6.588)	9(13.433)	8(10.979)
21-29	15(8.075)	14(16.467)	9(13.458)
30-38	9(14.663)	21(29.9)	39(24.438)
39-47	7(9.775)	22(19.933)	17(16.292)
48 +	6(11.9)	38(24.267)	12(19.833)

17. continued

$$\chi^2 = \frac{(14-6.588)^2}{6.588} + \frac{(9-13.433)^2}{13.433} + \frac{(8-10.979)^2}{10.979}$$

$$+ \frac{(15-8.075)^2}{8.075} + \frac{(14-16.467)^2}{16.467} + \frac{(9-13.458)^2}{13.458}$$

$$+ \frac{(9-14.663)^2}{14.663} + \frac{(21-29.9)^2}{29.9} + \frac{(39-24.438)^2}{24.438}$$

$$+ \frac{(7-9.775)^2}{9.775} + \frac{(22-19.933)^2}{19.933} + \frac{(17-16.292)^2}{16.292}$$

$$+ \frac{(6-11.9)^2}{11.9} + \frac{(38-24.267)^2}{24.267} + \frac{(12-19.833)^2}{19.833}$$

$$\chi^2 = 46.733$$

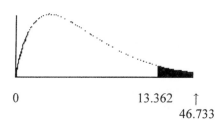

Reject the null hypothesis. There is enough evidence to support the claim that the type of movie selected is related to the age of the customer.

19.
H_0: The type of snack purchased is independent of the gender of the consumer. (claim)
H_1: The type of snack purchased is dependent upon the gender of the consumer.
C. V. = 4.605   d. f. = 2

Gender	Hot Dog	Peanuts	Popcorn	Total
Male	12(13.265)	21(15.388)	19(23.347)	52
Female	13(11.735)	8(13.612)	25(20.653)	46
Total	25	29	44	98

$$\chi^2 = \sum \frac{(O-E)^2}{E} = \frac{(12-13.265)^2}{13.265} + \frac{(21-15.388)^2}{15.388}$$

$$+ \frac{(19-23.347)^2}{23.347} + \frac{(13-11.735)^2}{11.735} + \frac{(8-13.612)^2}{13.612}$$

$$+ \frac{(25-20.653)^2}{20.653} = 6.342$$

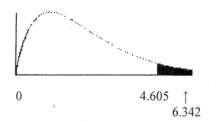

114

19. continued
Reject the null hypothesis. There is enough
evidence to reject the claim that the type of
snack chosen is independent of the gender of
the individual.

21.
H_0: The type of book selected by the
individual is independent of the gender of
the individual. (claim)
H_1: The type of book selected by the
individual is dependent on the gender of the
individual.
$\alpha = 0.05$ d. f. = 2

Gender	Mystery	Romance	Self-help	Total
Male	243(214.121)	201(198.260)	191(222.618)	635
Female	135(163.879)	149(151.740)	202(170.382)	486
Total	378	350	393	1121

$$\chi^2 = \sum \frac{(O-E)^2}{E} = \frac{(243-214.121)^2}{214.121} + \frac{(201-198.260)^2}{198.260}$$

$$+ \frac{(191-222.618)^2}{222.618} + \frac{(135-163.879)^2}{163.879}$$

$$+ \frac{(149-151.740)^2}{151.740} + \frac{(202-170.382)^2}{170.382} = 19.429$$

P-value < 0.005 (0.00006)
Reject the null hypothesis since P-value <
0.05. There is enough evidence to reject the
claim that the type of book purchased is
independent of gender.

23.
H_0: $p_1 = p_2 = p_3 = p_4$ (claim)
H_1: At least one proportion is different.

C. V. = 7.815   d. f. = 3

$E(passed) = \frac{120(167)}{120} = 41.75$

$E(failed) = \frac{120(313)}{120} = 78.25$

	Southside	West End	East Hills	Jefferson
Passed	49(41.75)	38(41.75)	46(41.75)	34(41.75)
Failed	71(78.25)	82(78.25)	74(78.25)	86(78.25)

$$\chi^2 = \frac{(49-41.75)^2}{41.75} + \frac{(38-41.75)^2}{41.75} + \frac{(46-41.75)^2}{41.75}$$

$$+ \frac{(34-41.75)^2}{41.75} + \frac{(71-78.25)^2}{78.25} + \frac{(82-78.25)^2}{78.25}$$

$$+ \frac{(74-78.25)^2}{78.25} + \frac{(86-78.25)^2}{78.25} = 5.317$$

23. continued

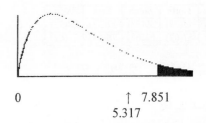

$$0 \qquad\qquad \underset{5.317}{\overset{\uparrow\ 7.851}{}}$$

Do not reject the null hypothesis. There is
not enough evidence to reject the claim that
the proportions are equal.

25.
H_0: $p_1 = p_2 = p_3 = p_4$ (claim)
H_1: At least one proportion is different.
C. V. = 7.815   d. f. = 3

$E(yes) = \frac{107(86)}{344} = 26.75$

$E(no) = \frac{237(86)}{344} = 59.25$

	21-29	30-39	40-49	50 +
Yes	32(26.75)	28(26.75)	26(26.75)	21(26.75)
No	54(59.25)	58(59.25)	60(59.25)	65(59.25)

$$\chi^2 = \frac{(32-26.75)^2}{26.75} + \frac{(28-26.75)^2}{26.75} + \frac{(26-26.75)^2}{26.75} +$$

$$\frac{(21-26.75)^2}{26.75} + \frac{(54-59.25)^2}{59.25} + \frac{(58-59.25)^2}{59.25} + \frac{(60-59.25)^2}{59.25}$$

$$+ \frac{(65-59.25)^2}{59.25} = 3.405$$

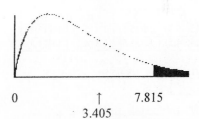

$$0 \qquad\qquad \underset{3.405}{\overset{\uparrow}{}} \qquad 7.815$$

Do not reject the null hypothesis. There is
not enough evidence to reject the claim that
the proportions are the same.

27.
H_0: $p_1 = p_2 = p_3 = p_4$ (claim)
H_1: At least one proportion is different.
C. V. = 6.251   d. f. = 3

$E(yes) = \frac{(100)(132)}{400} = 33$

$E(no) = \frac{(100)(268)}{400} = 67$

27. continued

	North	South	East	West
Yes	43(33)	39(33)	22(33)	28(33)
No	57(67)	61(67)	78(67)	72(67)

$$\chi^2 = \frac{(43-33)^2}{33} + \frac{(39-33)^2}{33} + \frac{(22-33)^2}{33} +$$

$$\frac{(28-33)^2}{33} + \frac{(57-67)^2}{67} + \frac{(61-67)^2}{67} + \frac{(78-67)^2}{67}$$

$$+ \frac{(72-67)^2}{67} = 12.755$$

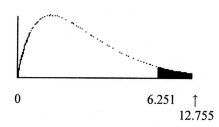

0   6.251   ↑
              12.755

Reject the null hypothesis. There is enough evidence to reject the claim that the proportions are the same.

29.
H_0: $p_1 = p_2 = p_3 = p_4$ (claim)
H_1: At least one proportion is different.
$\alpha = 0.05$ d. f. = 3

$$E(\text{on bars}) = \frac{30(62)}{120} = 15.5$$

$$E(\text{not on bars}) = \frac{30(58)}{120} = 14.5$$

	N	S	E	W
on	15(15.5)	18(15.5)	13(15.5)	16(15.5)
off	15(14.5)	12(14.5)	17(14.5)	14(14.5)

$$\chi^2 = \frac{(15-15.5)^2}{15.5} + \frac{(18-15.5)^2}{15.5} + \frac{(13-15.5)^2}{15.5} +$$

$$\frac{(16-15.5)^2}{15.5} + \frac{(15-14.5)^2}{14.5} + \frac{(12-14.5)^2}{14.5} +$$

$$\frac{(17-14.5)^2}{14.5} + \frac{(14-14.5)^2}{14.5} = 1.734$$

P-value > 0.10 (0.629)
Do not reject the null hypothesis. There is not enough evidence to reject the claim that the proportions are the same.

31.
H_0: $p_1 = p_2 = p_3$ (claim)
H_1: At least one proportion is different.
C. V. = 4.605   d. f. = 2

31. continued
$$E(\text{list}) = \frac{96(219)}{288} = 73$$

$$E(\text{no list}) = \frac{96(69)}{288} = 23$$

	A	B	C
list	77(73)	74(73)	68(73)
no list	19(23)	22((23)	28(23)

$$\chi^2 = \frac{(77-73)^2}{73} + \frac{(74-73)^2}{73} + \frac{(68-73)^2}{73}$$

$$+ \frac{(19-23)^2}{23} + \frac{(22-23)^2}{23} + \frac{(28-23)^2}{23}$$

$$\chi^2 = 2.401$$

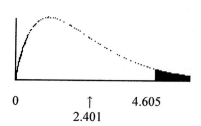

0   ↑   4.605
           2.401

Do not reject the null hypothesis. There is not enough evidence to reject the claim that the proportions are the same.

33.
$$\chi^2 = \frac{(|O-E|-0.5)^2}{E} = \frac{(|12-9.6|-0.5)^2}{9.6}$$

$$+ \frac{(|15-17.4|-0.5)^2}{17.4} + \frac{(|9-11.4|-0.5)^2}{11.4}$$

$$+ \frac{(|23-20.6|-0.5)^2}{20.6}$$

$$= \frac{3.61}{9.6} + \frac{3.61}{17.4} + \frac{3.61}{11.4} + \frac{3.61}{20.6}$$

$$= 0.376 + 0.207 + 0.317 + 0.175 = 1.075$$

REVIEW EXERCISES - CHAPTER 11

1.
H_0: The number of sales is equally distributed over five regions. (claim)
H_1: The null hypothesis is not true.

C. V. = 9.488   d. f. = 4
$E = \frac{1328}{5} = 265.6$

$$\chi^2 = \sum \frac{(O-E)^2}{E} = \frac{(236-265.6)^2}{265.6}$$

$$+ \frac{(324-265.6)^2}{265.6} + \frac{(182-265.6)^2}{265.6}$$

1. continued

$$+ \frac{(221-265.6)^2}{265.6} + \frac{(365-265.6)^2}{265.6} = 87.14$$

Alternate Solution:

O	E	O − E	(O − E)²	$\frac{(O-E)^2}{E}$
236	265.6	-29.6	876.18	3.299
324	265.6	58.4	3410.56	12.841
182	265.6	-83.6	6988.96	26.314
221	265.6	-44.6	1989.16	7.489
365	265.6	99.4	9880.36	37.200
				87.143

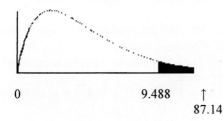

0   9.488   ↑
              87.14

Reject the null hypothesis. There is enough evidence to reject the claim that the number of items sold in each region is the same.

3.

H_0: The gender of the individual is not related to whether or not a person would use the labels. (claim)

H_1: The gender is related to use of the labels.

C. V. = 4.605   d. f. = 2

Gender	Yes	No	Undecided
Men	114(120.968)	30(22.258)	6(6.774)
Women	136(129.032)	16(23.742)	8(7.226)

$$\chi^2 = \frac{(114-120.968)^2}{120.968} + \frac{(30-22.258)^2}{22.258} + \frac{(6-6.774)^2}{6.774}$$

$$+ \frac{(136-129.032)^2}{129.032} + \frac{(16-23.742)^2}{23.742} + \frac{(8-7.226)^2}{7.226}$$

$$\chi^2 = 6.16$$

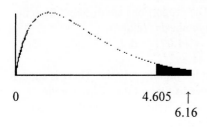

0   4.605   ↑
              6.16

3. continued

Reject the null hypothesis. There is not enough evidence to support the claim that opinion is independent of gender.

5.

H_0: The type of investment is independent of the age of the investor.

H_1: The type of investment is dependent upon the age of the investor. (claim)

C. V. = 9.488   d. f. = 4

Age	Large	Small	Inter.
45	20(20.18)	10(15.45)	10(15.45
65	42(33.82)	24(18.55)	24(18.55)

Age	CD	Bond
45	15(9.55)	45(31.36)
65	6(11.45)	24(37.64)

$$\chi^2 = \frac{(20-20.18)^2}{20.18} + \frac{(10-15.45)^2}{15.45} + \frac{(10-15.45)^2}{15.45}$$

$$+ \frac{(15-9.55)^2}{9.55} + \frac{(45-31.36)^2}{31.36} + \frac{(42-33.82)^2}{33.82} +$$

$$\frac{(24-18.55)^2}{18.55} + \frac{(24-18.55)^2}{18.55} + \frac{(6-11.45)^2}{11.45} +$$

$$\frac{(24-37.64)^2}{37.64} = 25.6$$

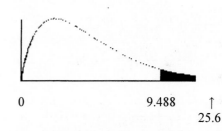

0   9.488   ↑
              25.6

Reject the null hypothesis. There is enough evidence to support the claim that the type of investment is dependent on age.

7.

H_0: $p_1 = p_2 = p_3$ (claim)

H_1: At least one proportion is different.

$\alpha = 0.01$ d. f. = 2

$$E(\text{work}) = \frac{80(114)}{240} = 38$$

$$E(\text{don't work}) = \frac{80(126)}{240} = 42$$

	16	17	18
work	45(38)	31(38)	38(38)
don't work	35(42)	49(42)	42(42)

7. continued

$$\chi^2 = \frac{(45-38)^2}{38} + \frac{(31-38)^2}{38} + \frac{(38-38)^2}{38}$$

$$+ \frac{(35-42)^2}{42} + \frac{(49-42)^2}{42} + \frac{(42-42)^2}{42} = 4.912$$

$0.05 < P\text{-value} < 0.10 \ (0.086)$

Do not reject the null hypothesis since P-value > 0.01. There is not enough evidence to reject the claim that the proportions are the same.

9.
$H_0: p_1 = p_2 = p_3 = p_4$
$H_1:$ At least one proportion is different.
C. V. = 6.251   d. f. = 3

$$E(\text{yes}) = \frac{50(58)}{200} = 14.5$$

$$E(\text{no}) = \frac{50(142)}{200} = 35.5$$

	A	B	C	D
Yes	12(14.5)	15(14.5)	10(14.5)	21(14.5)
No	38(35.5)	35(35.5)	40(35.5)	29(35.5)

$$\chi^2 = \frac{(12-14.5)^2}{14.5} + \frac{(15-14.5)^2}{14.5} + \frac{(10-14.5)^2}{14.5} +$$

$$\frac{(21-14.5)^2}{14.5} + \frac{(38-35.5)^2}{35.5} + \frac{(35-35.5)^2}{35.5}$$

$$+ \frac{(40-35.5)^2}{35.5} + \frac{(29-35.5)^2}{35.5} = 6.702$$

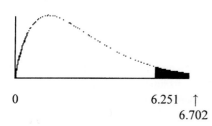

0   6.251 ↑
                                      6.702

Reject the null hypothesis. There is enough evidence to reject the claim that the proportions are the same.

CHAPTER 11 QUIZ

1. False, it is one-tailed right.
2. True
3. False, there is little agreement between observed and expected frequencies.
4. c.
5. b.
6. d.

7. 6
8. independent
9. right
10. at least five

11. H_0: The number of advertisements is equally distributed over five geographic regions.   (claim)
H_1: The number of advertisements is not equally distributed over five regions.
C. V. = 9.488   d. f. = 4   E = 240.4
$\chi^2 = \sum \frac{(O-E)^2}{E} = 45.4$
Reject the null hypothesis. There is enough evidence to reject the claim that the number of advertisements is equally distributed.

12. H_0: The ads produced the same number of responses. (claim)
H_1: The ads produced different numbers of responses.
C. V. = 13.277   d. f. = 4   E = 64.6
$\chi^2 = \sum \frac{(O-E)^2}{E} = 12.6$
Do not reject the null hypothesis. There is not enough evidence to reject the claim that the ads produced the same number of responses.

13. H_0: 62% of respondents never watch shopping channels, 23% watch rarely, 11% watch occasionally, and 4% watch frequently.
H_1: College students show a different preference for shopping channels. (claim)
C. V. = 7.815   d. f. = 3
$\chi^2 = 21.789$
Reject the null hypothesis. There is enough evidence to support the claim that college students show a different preference for shopping channels.

14. H_0: The number of commuters is distributed as follows: alone - 75.7%, carpooling - 12.2%, public transportation - 4.7%, walking - 2.9%, other - 1.2%, and working at home - 3.3%.
H_1: The proportions are different from the null hypothesis. (claim)
C. V. = 11.071   d. f. = 5
$\chi^2 = 41.692$
Reject the null hypothesis. There is enough evidence to support the claim that the distribution is different from the one stated in the null hypothesis.

15. H_0: The type of novel purchased is independent of the gender of the purchaser. (claim)
H_1: The type of novel purchased is dependent on the gender of the purchaser.
C. V. = 5.991   d. f. = 2
$\chi^2 = 132.9$
Reject the null hypothesis. There is enough evidence to reject the claim that the novel purchased is independent of the gender of the purchaser.

16. H_0: The type of pizza ordered is independent of the age of the purchaser. (claim)
H_1: The type of pizza ordered is dependent on the age of the purchaser.
$\alpha = 0.10$ d. f. = 9
$\chi^2 = 107.3$
P-value < 0.005
Reject the null hypothesis since P-value < 0.10. There is enough evidence to reject the claim that the type of pizza is independent of the age of the purchaser.

17. H_0: The color of the pennant purchased is independent of the gender of the purchaser. (claim)
H_1: The color of the pennant purchased is dependent on the gender of the purchaser.
C. V. = 4.605   d. f. = 2
$\chi^2 = 5.6$
Reject the null hypothesis. There is enough evidence to reject the claim that the color of the pennant purchased is independent of the gender of the purchaser.

18. H_0: The opinion of the children on the use of the tax credit is independent of the gender of the children.
H_1: The opinion of the children on the use of the tax credit is dependent upon the gender of the children. (claim)
C. V. = 4.605   d. f. = 2
$\chi^2 = 1.534$
Do not reject the null hypothesis. There is not enough evidence to support the claim that the opinion of the children is dependent upon their gender.

Note: Graphs are not to scale and are intended to convey a general idea. Answers may vary due to rounding.

EXERCISE SET 12-3

1.
The analysis of variance using the F-test can be used to compare 3 or more means.

3.
The populations from which the samples were obtained must be normally distributed. The samples must be independent of each other. The variances of the populations must be equal.

5.
$$F = \frac{s_B^2}{s_W^2}$$

7.
Scheffe′ Test and the Tukey Test

9.
$H_0: \mu_1 = \mu_2 = \mu_3$
$H_1:$ At least one mean is different from the others. (claim)
C. V. = 3.47 $\alpha = 0.05$
d. f. N. = 2   d. f. D. = 21

$\overline{X}_{GM} = 4.554$ $s_B^2 = 9.82113$ $s_W^2 = 4.93225$

$$F = \frac{9.82113}{4.93225} = 1.9912$$

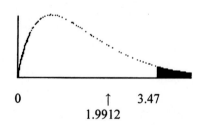

0   ↑   3.47
                   1.9912

Do not reject the null hypothesis. There is not enough evidence to support the claim that at least one mean is different from the others.

11.
$H_0: \mu_1 = \mu_2 = \mu_3$
$H_1:$ At least one mean is different from the others. (claim)
C. V. = 3.98 $\alpha = 0.05$
d. f. N = 2   d. f. D = 11

11. continued
$$\overline{X}_{GM} = \frac{52414}{14} = 3743.857$$

$$s_B^2 = 3,633,540.88$$

$$s_W^2 = 1,330,350$$

$$F = \frac{3633540.88}{1330350} = 2.7313$$

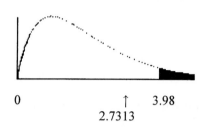

0   ↑   3.98
                  2.7313

Do not reject the null hypothesis. There is not enough evidence to support the claim that at least one mean is different from the others.

13.
$H_0: \mu_1 = \mu_2 = \mu_3$ (claim)
$H_1:$ At least one mean is different.
k = 3   N = 17   d.f.N. = 2   d.f.D. = 14
CV = 3.74

$\overline{X}_1 = 15,637$ $s_1^2 = 1,748,779.5$
$\overline{X}_2 = 14,254.5$ $s_2^2 = 13,830.7$
$\overline{X}_3 = 14,216.33$ $s_3^2 = 30,730.67$

$$\overline{X}_{GM} = 14,647.647$$

$$s_B^2 = \frac{5(15637 - 14647.647)^2}{2} + \frac{6(14254.5 - 14647.647)^2}{2}$$

$$+ \frac{6(14216.3333 - 14647.647)^2}{2} = 3,468,836.61$$

$$s_W^2 = \frac{4(1748779.5) + 5(13830.7) + 5(30730.667)}{4 + 5 + 5}$$

$$s_W^2 = 515,566.06$$

$$F = \frac{3468836.61}{515566.06} = 6.7282$$

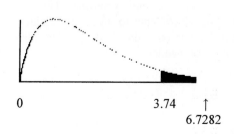

0   3.74   ↑
                           6.7282

13. continued
Reject the null hypothesis. There is not enough evidence to support the claim that the means are the same.

Scheffe Test:
C. V. = 7.48

$$F_s = \frac{(\bar{X}_i - \bar{X}_j)^2}{s_W^2(\frac{1}{n_i} + \frac{1}{n_j})}$$

For $\bar{X}_1$ vs $\bar{X}_2$

$$F_S = \frac{(15637 - 14254.5)^2}{515566.06(\frac{1}{5} + \frac{1}{6})} = 10.111$$

For $\bar{X}_1$ vs $\bar{X}_3$

$$F_S = \frac{(15637 - 14216.33)^2}{515566.06(\frac{1}{5} + \frac{1}{6})} = 10.677$$

For $\bar{X}_2$ vs $\bar{X}_3$

$$F_S = \frac{(14254.5 - 14216.33)^2}{515566.06(\frac{1}{6} + \frac{1}{6})} = 0.0085$$

There is a significant difference between $\bar{X}_1$ and $\bar{X}_2$ and $\bar{X}_1$ and $\bar{X}_3$.

15.
H_0: $\mu_1 = \mu_2 = \mu_3$ (claim)
H_1: At least one mean is different from the others.
C. V. = 4.10 $\alpha = 0.10$
d. f. N = 2   d. f. D = 10

$\bar{X}_1 = 35.4$ $s_1^2 = 351.8$
$\bar{X}_2 = 68.75$ $s_2^2 = 338.25$
$\bar{X}_3 = 44.25$ $s_3^2 = 277.583$

$\bar{X}_{GM} = \frac{629}{13} = 48.385$

$$s_B^2 = \frac{\sum n_i(\bar{X}_i - \bar{X}_{GM})^2}{k-1}$$

$$s_B^2 = \frac{5(35.4 - 48.385)^2}{2} + \frac{4(68.75 - 48.385)^2}{2}$$

$$+ \frac{4(44.25 - 48.385)^2}{2} = 1285.188$$

$$s_W^2 = \frac{\sum(n_i - 1)s_i^2}{\sum(n_i - 1)}$$

$$= \frac{4(351.8) + 3(338.25) + 3(277.583)}{4 + 3 + 3}$$

$$= 325.47$$

15. continued
$$F = \frac{s_B^2}{s_W^2} = \frac{1285.188}{325.47} = 3.9487$$

Do not reject the null hypothesis. There is enough evidence to support the claim that the means are the same.

17.
H_0: $\mu_1 = \mu_2 = \mu_3$
H_1: At least one mean is different from the others. (claim)
C. V. = 2.61 $\alpha = 0.10$
d. f. N = 2   d. f. D = 19

$\bar{X}_1 = 233.33$ $s_1 = 28.225$
$\bar{X}_2 = 203.125$ $s_2 = 39.364$
$\bar{X}_3 = 155.625$ $s_3 = 28.213$

$\bar{X}_{GM} = 194.091$

$$s_B^2 = \frac{21,729.735}{2} = 10,864.8675$$

$$s_W^2 = \frac{20,402.083}{19} = 1073.794$$

$$F = \frac{s_B^2}{s_W^2} = \frac{10,864.8675}{1073.794} = 10.12$$

P-value = 0.00102
Reject since P-value < 0.10.

Scheffe Test   C. V. = 5.24

$\bar{X}_1$ vs $\bar{X}_2$:

$$F_s = \frac{(\bar{X}_i - \bar{X}_j)^2}{s_W^2(\frac{1}{n_i} + \frac{1}{n_j})} = \frac{(233.33 - 203.125)^2}{1073.776(\frac{1}{6} + \frac{1}{8})}$$

$$F_s = 2.91$$

$\bar{X}_1$ vs $\bar{X}_3$:

$$F_s = \frac{(233.33 - 155.625)^2}{1073.776(\frac{1}{6} + \frac{1}{8})} = 19.28$$

$\bar{X}_2$ vs $\bar{X}_3$:

$$F_s = \frac{(203.125 - 155.625)^2}{1073.776(\frac{1}{8} + \frac{1}{8})} = 8.40$$

There is a significant difference between $\bar{X}_1$ and $\bar{X}_3$ and between $\bar{X}_2$ and $\bar{X}_3$.

19.
H_0: $\mu_1 = \mu_2 = \mu_3 = \mu_4$
H_1: At least one mean is different. (claim)

19. continued
C. V. = 4.13 $\alpha = 0.01$
d. f. N = 3   d. f. D = 64

$$\overline{X}_{GM} = \frac{36,254}{68} = 533.147$$

$$s_B^2 = \frac{40,968.145}{3} = 13,656.05$$

$$s_w^2 = \frac{787,506.385}{64} = 12,304.787$$

$$F = \frac{13,656.05}{12,304.787} = 1.11$$

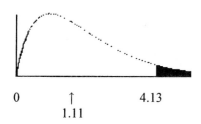

```
0          ↑              4.13
         1.11
```

Do not reject the null hypothesis. There is not enough evidence to support the claim that at least one mean is different.

EXERCISE SET 12-4

1.
The two-way ANOVA allows the researcher to test the effects of two independent variables and a possible interaction effect. The one-way ANOVA can test the effects of one independent variable only.

3.
The mean square values are computed by dividing the sum of squares by the corresponding degrees of freedom.

5.
a. d. f.$_A$ = (3 − 1) = 2 for factor A
b. d. f.$_B$ = (2 − 1) = 1 for factor B
c. d. f.$_{A \times B}$ = (3 − 1)(2 − 1) = 2
d. d. f.$_{within}$ = 3 · 2(5 − 1) = 24

7.
The two types of interactions that can occur are ordinal and disordinal.

9.
a. The lines will be parallel or approximately parallel. They could also coincide.

9. continued
b. The lines will not intersect and they will not be parallel.
c. The lines will intersect.

11.
H_0: There is no interaction effect between the time of day and the type of diet on a person's sodium level.
H_1: There is an interaction effect between the time of day and the type of diet on a person's sodium level.

H_0: There is no difference between the means for the sodium level for the times of day.
H_1: There is a difference between the means for the sodium level for the times of day.

H_0: There is no difference between the means for the sodium level for the type of diet.
H_1: There is a difference between the means for the sodium level for the type of diet.

ANOVA SUMMARY TABLE

Source	SS	d. f.	MS	F
Time	1800	1	1800	25.806
Diet	242	1	242	3.470
Interaction	264.5	1	264.5	3.792
Within	279	4	69.75	
Total	2585.5	7		

The critical value at $\alpha = 0.05$ with d. f. N = 1 and d. f. D = 4 is 7.71 for F_A, F_B, and $F_{A \times B}$.

Since the only F test value that exceeds 7.71 is the one for the time, 25.806, it can be concluded that there is a difference in the means for the sodium level taken at two different times.

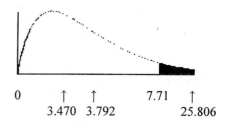

```
0      ↑    ↑        7.71    ↑
     3.470 3.792          25.806
```

13.
H_0: There is no interaction effect between the gender of the individual and the duration of the training on the test scores.

13. continued

H_1: There is an interaction effect between the gender of the individual and the duration of the training on the test scores.

H_0: There is no difference between the means of the test scores for the males and females.

H_1: There is a difference between the means of the test scores for the males and females.

H_0: There is no difference between the means of the test scores for the two different durations.

H_1: There is a difference between the means of the test scores for the two different durations.

ANOVA SUMMARY TABLE

Source	SS	d. f.	MS	F
Gender	57.042	1	57.042	0.835
Duration	7.042	1	7.042	0.103
Interaction	3978.375	1	3978.375	58.270
Within	1365.5	20	68.275	
Total	5407.959	23		

The critical value at $\alpha = 0.10$ with d. f. N = 1 and d. f. D = 20 is 2.97. Since the F test value for the interaction is greater than the critical value, it can be concluded that the gender affects the test scores differently for the duration levels.

15.

H_0: There is no interaction effect between the ages of the salespersons and the products they sell on the monthly sales.

H_1: There is an interaction effect between the ages of the salespersons and the products they sell on the monthly sales.

H_0: There is no difference in the means of the monthly sales of the two age groups.

H_1: There is a difference in the means of the monthly sales of the two age groups.

H_0: There is no difference among the means of the sales for the different products.

H_1: There is a difference among the means of the sales for the different products.

15. continued

ANOVA SUMMARY TABLE

Source	SS	d. f.	MS	F
Age	168.033	1	168.033	1.567
Product	1762.067	2	881.034	8.215
Interaction	7955.267	2	3877.634	37.087
Error	2574	24	107.250	
Total	12459.367	29		

At $\alpha = 0.05$, the critical values are:

For age, d. f. N = 1, d. f. D = 24, C. V. = 4.26

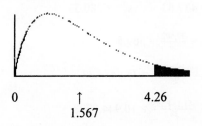

0   ↑   4.26
        1.567

For product and interaction, d. f. N = 2, d. f. D = 24, and C. V. = 3.40

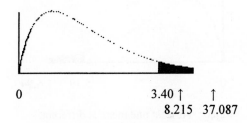

0   3.40 ↑   ↑
                        8.215 37.087

The null hypotheses for the interaction effect and for the type of product sold are rejected since the F test values exceed the critical value, 3.40. The cell means are:

Age	Pools	Spas	Saunas
over 30	38.8	28.6	55.4
30 & under	21.2	68.6	18.8

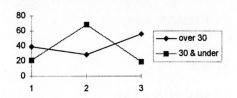

Since the lines cross, there is a disordinal interaction hence there is an interaction effect between the age of the sales person and the type of products sold on the sales.

REVIEW EXERCISES - CHAPTER 12

1.
H_0: $\mu_1 = \mu_2 = \mu_3$ (claim)
H_1: At least one mean is different from the others.
C. V. = 5.39 $\alpha = 0.01$
d. f. N = 2   d. f. D = 33

$\overline{X}_1 = 620.5$ $s_1^2 = 5445.91$

$\overline{X}_2 = 610.17$ $s_2^2 = 22{,}108.7$

$\overline{X}_3 = 477.83$ $s_3^2 = 5280.33$

$\overline{X}_{GM} = \frac{20{,}502}{36} = 569.5$

$s_B^2 = \frac{151{,}890.667}{2} = 75{,}945.333$

$s_W^2 = \frac{361{,}184.333}{33} = 10{,}944.9798$

$F = \frac{s_B^2}{s_W^2} = \frac{75{,}945.333}{10{,}944.9798} = 6.94$

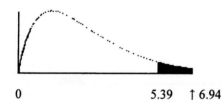

0   5.39   ↑ 6.94

Reject. At least one mean is different.

Tukey Test   C. V. = 4.45 using (3, 33)

$\overline{X}_1$ vs $\overline{X}_2$

$q = \frac{\overline{X}_1 - \overline{X}_2}{\sqrt{\frac{s_W^2}{n}}} = \frac{620.5 - 610.17}{\sqrt{\frac{10{,}944.98}{12}}} = 0.342$

$\overline{X}_1$ vs $\overline{X}_3$

$q = \frac{620.5 - 477.83}{\sqrt{\frac{10{,}944.98}{12}}} = 4.72$

$\overline{X}_2$ vs $\overline{X}_3$

$q = \frac{610.17 - 477.83}{\sqrt{\frac{10{,}944.98}{12}}} = 4.38$

There is a significant difference between $\overline{X}_1$ and $\overline{X}_3$.

3.
H_0: $\mu_1 = \mu_2 = \mu_3$
H_1: At least one mean is different from the others. (claim)

C. V. = 3.55 $\alpha = 0.05$
d. f. N = 2   d. f. D = 18

$\overline{X}_1 = 29.625$ $s_1^2 = 59.125$

$\overline{X}_2 = 29$ $s_2^2 = 63.333$

$\overline{X}_3 = 28.5$ $s_3^2 = 37.1$

$\overline{X}_{GM} = 29.095$

$s_B^2 = \frac{\sum n_i (\overline{X}_i - \overline{X}_{GM})^2}{k-1}$

$s_B^2 = \frac{8(29.625 - 29.095)^2}{2} + \frac{7(29 - 29.095)^2}{2}$

$+ \frac{6(28.5 - 29.095)^2}{2} = 2.21726$

$s_W^2 = \frac{\sum(n_i - 1)s_i^2}{\sum(n_i - 1)}$

$s_W^2 = \frac{7(59.125) + 6(63.333) + 5(37.1)}{7 + 6 + 5}$

$s_W^2 = 54.509611$

$F = \frac{s_B^2}{s_W^2} = \frac{2.21726}{54.509611} = 0.04075$

Do not reject the null hypothesis. There is not enough evidence to support the claim that at least one mean is different from the others.

5.
H_0: $\mu_1 = \mu_2 = \mu_3$
H_1: At least one mean is different. (claim)
C. V. = 2.61 $\alpha = 0.10$
d. f. N = 2   d. f. D = 19

$\overline{X}_{GM} = 3.8591$

$s_B^2 = 1.65936$

$s_W^2 = 3.40287$

$F = \frac{1.65936}{3.40287} = 0.4876$

5. continued

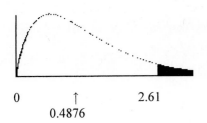

0   ↑   2.61
        0.4876

Do not reject. There is not enough evidence to support the claim that at least one mean is different from the others.

7.
H_0: $\mu_1 = \mu_2 = \mu_3$
H_1: At least one mean is different from the others.
C. V. = 3.68 $\alpha = 0.05$
d. f. N = 2   d. f. D = 15

$\overline{X}_{GM} = 34.611$

$s_B^2 = \frac{1445.7778}{2} = 722.8889$

$s_W^2 = \frac{1376.5}{15} = 91.7667$

$F = \frac{722.8889}{91.7667} = 7.8775$

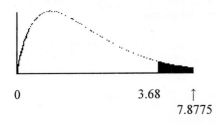

0   3.68   ↑
             7.8775

Reject the null hypothesis. There is enough evidence to support the claim that at least one mean is different from the others.

Tukey Test:
C. V. = 3.67

$\overline{X}_1$ vs $\overline{X}_2$:

$q = \frac{47.167 - 26.833}{\sqrt{\frac{91.767}{6}}} = 5.1994$

$\overline{X}_1$ vs $\overline{X}_3$:

$q = \frac{47.167 - 29.833}{\sqrt{\frac{91.767}{6}}} = 4.4323$

7. continued

$\overline{X}_2$ vs $\overline{X}_3$:

$q = \frac{26.833 - 29.833}{\sqrt{\frac{91.767}{6}}} = -0.7671$

There is a significant different between $\overline{X}_1$ and $\overline{X}_2$ and between $\overline{X}_1$ and $\overline{X}_3$.

9.
H_0: There is no interaction effect between the type of exercise program and the type of diet on a person's glucose level.
H_1: There is an interaction effect between the type of exercise program and the type of diet on a person's glucose level.

H_0: There is no difference in the means for the glucose levels of the persons in the two exercise programs.
H_1: There is a difference in the means for the glucose levels of the persons in the two exercise programs.

H_0: There is no difference in the means for the glucose levels of the persons in the two diet programs.
H_1: There is a difference in the means for the glucose levels of the persons in the two diet programs.

ANOVA SUMMARY TABLE

Source	SS	d. f.	MS	F
Exercise	816.75	1	816.75	60.5
Diet	102.083	1	102.083	7.562
Interaction	444.083	1	444.083	32.895
Within	108	8	13.5	
Total	1470.916	11		

At $\alpha = 0.05$ and d. f. N = 1 and d. f. D = 8 the critical value is 5.32 for each F_A, F_B, and $F_{A \times B}$.

Hence all three null hypotheses are rejected.

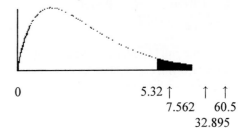

0   5.32 ↑   ↑   ↑
                      7.562   60.5
                         32.895

9. continued
The cell means should be calculated.

Exercise	Diet	
	A	B
I	64	57.667
II	68.333	86.333

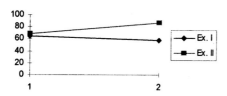

Since the means for the Exercise Program I are both smaller than those for Exercise Program II and the vertical differences are not the same, the interaction is ordinal. Hence one can say that there is a difference for exercise, diet; and that an interaction effect is present.

CHAPTER 12 QUIZ

1. False, there could be a significant difference between only some of the means.
2. False, degrees of freedom are used to find the critical value.
3. False, the null hypothesis should not be rejected.
4. True
5. d.
6. a.
7. a.
8. c.
9. ANOVA
10. Tukey
11. two

12. H_0: $\mu_1 = \mu_2 = \mu_3$ (claim)
H_1: At least one mean is different from the others.
C. V. = 3.55

$s_B^2 = 6.0476$ $s_W^2 = 0.8730$
$F = \frac{6.0476}{0.8730} = 6.927$

Reject the null hypothesis. There is not enough evidence to support the claim that the means are the same.

$\overline{X}_1 = 3.29$
$\overline{X}_2 = 5.14$

12. continued
$\overline{X}_3 = 4.29$

Tukey Test   C. V. = 3.61
$\overline{X}_1$ vs $\overline{X}_2$:

$q = \frac{3.29 - 5.14}{\sqrt{\frac{0.873}{7}}} = -5.24$

$\overline{X}_1$ vs $\overline{X}_3$:

$q = \frac{3.29 - 4.29}{\sqrt{\frac{0.873}{7}}} = -2.83$

$\overline{X}_2$ vs $\overline{X}_3$:

$q = \frac{5.14 - 4.29}{\sqrt{\frac{0.873}{7}}} = 2.41$

There is a significant difference between $\overline{X}_1$ and $\overline{X}_2$.

13. H_0: $\mu_1 = \mu_2 = \mu_3$ (claim)
H_1: At least one mean is different from the others.
C. V. = 6.01 $\alpha = 0.01$
$s_B^2 = 70.776$ $s_W^2 = 6.094$
$F = \frac{70.776}{6.094} = 11.614$

Reject the null hypothesis. There is not enough evidence to support the claim that the means are the same.

Scheffe Test   C. V. = 12.02
$\overline{X}_1$ vs $\overline{X}_2$:

$F_s = \frac{(27.75-26.667)^2}{6.094(\frac{1}{8}+\frac{1}{6})} = 0.660$

$\overline{X}_1$ vs $\overline{X}_3$:

$F_s = \frac{(27.75-21.857)^2}{6.094(\frac{1}{8}+\frac{1}{7})} = 21.274$

$\overline{X}_2$ vs $\overline{X}_3$:

$F_s = \frac{(26.667-21.857)^2}{6.094(\frac{1}{6}+\frac{1}{7})} = 12.266$

There is a significant difference between means 1 and 3 and means 2 and 3.

14. H_0: $\mu_1 = \mu_2 = \mu_3$
H_1: At least one mean is different from the others. (claim)
C. V. = 3.68 $\alpha = 0.05$
$s_B^2 = 617.167$ $s_W^2 = 58.811$

14. continued

$F = \frac{617.167}{58.811} = 10.494$

Reject the null hypothesis. There is enough evidence to support the claim that at least one mean is different from the others.

Tukey Test:
C. V. = 3.67
$\overline{X}_1 = 47.67$
$\overline{X}_2 = 63$
$\overline{X}_3 = 43.83$

$\overline{X}_1$ vs $\overline{X}_2$:

$q = \frac{47.67-63}{\sqrt{\frac{58.811}{6}}} = -4.90$

$\overline{X}_1$ vs $\overline{X}_3$:

$q = \frac{47.67-43.83}{\sqrt{\frac{58.811}{6}}} = 1.23$

$\overline{X}_2$ vs $\overline{X}_3$:

$q = \frac{63-43.83}{\sqrt{\frac{58.811}{6}}} = 6.12$

There is a significant difference between $\overline{X}_1$ and $\overline{X}_2$ and between $\overline{X}_2$ and $\overline{X}_3$.

15. H_0: $\mu_1 = \mu_2 = \mu_3$ (claim)
H_1: At least one mean is different from the others.
C. V. = 2.70 $\alpha = 0.10$
$s_B^2 = 0.1213$ $s_W^2 = 2.3836$
$F = \frac{0.1213}{2.3836} = 0.0509$
Do not reject. There is not enough evidence to reject the claim that the means are the same.

16. H_0: $\mu_1 = \mu_2 = \mu_3 = \mu_4$
H_1: At least one mean is different from the others. (claim)
C. V. = 3.07 $\alpha = 0.05$
$s_B^2 = 15.3016$ $s_W^2 = 33.5283$
$F = \frac{15.3016}{33.5283} = 0.4564$
Do not reject. There is not enough evidence to support the claim that at least one mean is different.

17.
a. two-way ANOVA
b. diet and exercise program
c. 2

17. continued
d. H_0: There is no interaction effect between the type of exercise program and the type of diet on a person's weight loss.
H_1: There is an interaction effect between the type of exercise program and the type of diet on a person's weight loss.

H_0: There is no difference in the means of the weight losses for those in the exercise programs.
H_1: There is a difference in the means of the weight losses for those in the exercise programs.

H_0: There is no difference in the means of the weight losses for those in the diet programs.
H_1: There is a difference in the means of the weight losses for those in the diet programs.

e. Diet: F = 21, significant
Exercise Program: F = 0.429, not significant
Interaction: F = 0.429, not significant

f. Reject the null hypothesis for the diets.

Note: Graphs are not to scale and are intended to convey a general idea. Answers may vary due to rounding.

EXERCISE SET 13-2

1.
Non-parametric means hypotheses other than those using population parameters can be tested, whereas distribution free means no assumptions about the population distributions have to be satisfied.

3.
The advantages of non-parametric methods are:
1. They can be used to test population parameters when the variable is not normally distributed.
2. They can be used when data is nominal or ordinal in nature.
3. They can be used to test hypotheses other than those involving population parameters.
4. The computations are easier in some cases than the computations of the parametric counterparts.
5. They are easier to understand.
The disadvantages are:
1. They are less sensitive than their parametric counterparts.
2. They tend to use less information than their parametric counterparts.
3. They are less efficient than their parametric counterparts.

5.

DATA	21	31	34	41	41	61	65	72
RANK	1	2	3	4.5	4.5	6	7	8

7.

DATA	3	5	5	6	7	8	8	9	12	14	15	17
RANK	1	2.5	2.5	4	5	6.5	6.5	8	9	10	11	12

9.

DATA	187	190	190	236	321	532	673
RANK	1	2.5	2.5	4	5	6	7

EXERCISE SET 13-3

1.
The sign test uses only + or − signs.

3.
The smaller number of + or − signs.

5.

```
+   −   −   −   −
+   +   +   +   +
−   −   +   +   0
−   −   +   +   +
```

H_0: Median = 2.8   (claim)
H_1: Median $\neq$ 2.8

$\alpha = 0.05$ $n = 19$
C. V. = 4
Test value = 8

$$\overset{8\,\downarrow}{\underset{3\quad 4\quad 5\quad 6\quad 7\quad\quad 8\quad 9}{|\quad|\quad|\quad|\quad|\quad\quad|\quad|}}$$

Do not reject the null hypothesis. There is not enough evidence to reject the claim that the median is 2.8.

7.

```
+   +   +   +
−   +   +   +
+   −   −   +
```

H_0: Median = \$325   (claim)
H_1: Median $\neq$ \$325

$\alpha = 0.05$ $n = 12$
C. V. = 2   Test value = 3

$$\overset{3\,\downarrow}{\underset{0\quad 1\quad 2\quad\quad 3\quad\quad 4}{|\quad|\quad|\quad\quad|\quad\quad|}}$$

Do not reject the null hypothesis. There is not enough evidence to reject the claim that the median rent is \$325.

9.
H_0: median = 7.30   (claim)
H_1: median $\neq$ 7.30

C. V. = ± 1.96

$$z = \frac{(x+0.5)-\left(\frac{n}{2}\right)}{\frac{\sqrt{n}}{2}} = \frac{(15+0.5)-\frac{37}{2}}{\frac{\sqrt{37}}{2}}$$

$$= \frac{-3}{3.041} = -0.99$$

9. continued

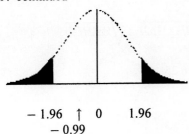

$-1.96 \quad \uparrow \quad 0 \quad 1.96$
-0.99

Do not reject the null hypothesis. There is not enough evidence to reject the claim that the median is 7.30 minutes.

11.
H_0: $p \leq 50\%$
H_1: $p > 50\%$ (claim)

$$z = \frac{(x + 0.5) - \left(\frac{n}{2}\right)}{\frac{\sqrt{n}}{2}} = \frac{(21 + 0.5) - \frac{50}{2}}{\frac{\sqrt{50}}{2}}$$

$$= \frac{21.5 - 25}{3.54} = \frac{-3.5}{3.54} = -0.99$$

P-value = 0.1611

Do not reject. There is not enough evidence to support the claim that more than 50% of the students favor single room dormitories.

13.
H_0: Median = 50   (claim)
H_1: Median $\neq$ 50

$$z = \frac{(x + 0.5) - \left(\frac{n}{2}\right)}{\frac{\sqrt{n}}{2}} = \frac{(38 + 0.5) - \frac{100}{2}}{\frac{\sqrt{100}}{2}}$$

$$= \frac{38.5 - 50}{5} = \frac{-13.5}{5} = -2.3$$

P-value = 0.0057

Reject. There is enough evidence to reject the claim that 50% of the students are against extending the school year.

15.

A	B	C	D	E	F	G	H
+	+	+	+	+	−	+	+

H_0: The medication has no effect on weight loss.
H_1: The medication affects weight loss. (claim)
$\alpha = 0.05$ n = 8

15. continued
C. V. = 0   Test value = 1

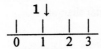

1 ↓
0 1 2 3

Do not reject the null hypothesis. There is not enough evidence to support the claim that the medication affects weight loss.

17.

1	2	3	4	5	6	7	8	9
−	−	+	−	−	−	−	−	−

H_0: Reasoning ability will not be affected by the course.
H_1: Reasoning ability increased after the course.   (claim)
$\alpha = 0.05$ n = 9
C. V. = 1   Test value = 1

1 ↓
0 1 2 3

Reject the null hypothesis. There is enough evidence to support the claim that the reasoning ability has increased after the course.

19.

1	2	3	4	5	6	7	8	9	10
−	+	+	+	+	+	−	0	−	+

H_0: Alcohol has no effect on a person's I. Q. test score.
H_1: Alcohol does effect a person's I. Q. Test score.   (claim)

$\alpha = 0.10$ n = 9
C. V. = 1   Test value = 3

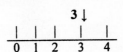

3 ↓
0 1 2 3 4

Do not reject the null hypothesis. There is not enough evidence to reject the claim that alcohol has no effect on a person's I. Q. score.

21.
3, 4, 6, 9, 12, 15, 15, 16, 18, 22, 25, 30

21. continued
At $\alpha = 0.05$, the value from Table J with $n = 12$ is 2; hence, count in 3 numbers from each end to get $6 \leq MD \leq 22$.

23.
4.2, 4.5, 4.7, 4.8, 5.1, 5.2, 5.6, 6.3, 7.1, 7.2, 7.8, 8.2, 9.3, 9.3, 9.5, 9.6
At $\alpha = 0.02$, the value from Table J with $n = 16$ is 2; hence, count 3 numbers from each end to get $4.7 \leq MD \leq 9.3$

25.
12, 14, 14, 15, 16, 17, 18, 19, 19, 21, 23, 25, 27, 32, 33, 35, 39, 41, 42, 47
At $\alpha = 0.05$, the value from Table J with $n = 20$ is 5; hence, count in 6 numbers from each end to get $17 \leq MD \leq 33$.

EXERCISE SET 13-4

1.
The sample sizes must be greater than or equal to 10.

3.
The standard normal distribution.

5.
H_0: There is no difference in the number of books each group read. (claim)
H_1: There is a difference in the number of books each group read.
C. V. $= \pm 1.65$

0	1	2	3	3	4	4	5	5
1	2	3	4.5	4.5	6.5	6.5	8.5	8.5
S	S	S	S	S	M	S	S	S

6	7	7	8	9	10	11	11	12
10	11.5	11.5	13	14	15	16.5	16.5	18.5
M	M	M	M	M	M	S	S	M

12	13	15	16	18
18.5	20	21	22	23
S	M	M	S	M

$R = 164$

$$\mu_R = \frac{n_1(n_1 + n_2 + 1)}{2}$$

$$= \frac{11(11 + 12 + 1)}{2} = \frac{11(24)}{2} = \frac{264}{2} = 132$$

$$\sigma_R = \sqrt{\frac{n_1 \cdot n_2(n_1 + n_2 + 1)}{12}}$$

5. continued

$$\sigma_R = \sqrt{\frac{11 \cdot 12(11 + 12 + 1)}{12}} = \sqrt{\frac{(11)(12)(24)}{12}}$$

$$= \sqrt{264} = 16.25$$

$$Z = \frac{R - \mu_R}{\sigma_R} = \frac{164 - 132}{16.25} = 1.97$$

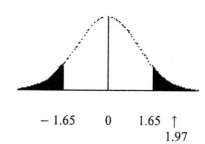

$$-1.65 \qquad 0 \qquad 1.65 \uparrow$$
$$1.97$$

Reject the null hypothesis. There is enough evidence to reject the claim that there is no difference in the number of books read by each group.

7.
H_0: There is no difference in the number of scholarships for the two high schools.
H_1: There is a difference between the number of scholarships for the two schools. (claim)

C. V. $= \pm 1.96$

1	2	3	3	3	4	4	4	4	5	5	6
1	2	4	4	4	7.5	7.5	7.5	7.5	10.5	10.5	13
V	O	V	O	O	V	V	O	V	O	O	V

6	6	7	7	7	8	8	9	9	10	11	12
13	13	16	16	16	18.5	18.5	20.5	20.5	22	23	24
O	O	V	V	O	V	O	V	O	V	V	V

$R = 156.5$ for Valley View High School
$R = 143.5$ for Ocean View High School

$$\mu_R = \frac{12(12 + 12 + 1)}{2} = 150$$

$$\sigma_R = \sqrt{\frac{12 \cdot 12(12 + 12 + 1)}{12}} = 17.32$$

$$z = \frac{156.5 - 150}{17.32} = 0.38$$

7. continued

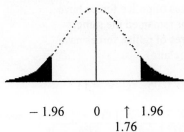

$-1.96 \quad 0 \quad \uparrow \ 1.96$
1.76

Do not reject the null hypothesis. There is enough evidence to support the claim that there is a difference in the number of scholarships.

9.
H_0: There is no difference in the amount of money awarded to each city.
H_1: There is a difference in the amount of money awarded to each city. (claim)
C. V. $= \pm 1.96$

232	239	324	431	453	563	601	602
1	2	3	4	5	6	7	8
A	B	B	A	B	A	A	A

605	626	648	687	718	752	769	824
9	10	11	12	13	14	15	16
A	B	A	A	B	B	B	A

832	869	885	918	921	925	927	953
17	18	19	20	21	22	23	24
B	B	B	B	A	A	B	A

R = 141   (for city A)

$$\mu_R = \frac{n_1(n_1 + n_2 + 1)}{2} = \frac{12(12 + 12 + 1)}{2} = 150$$

$$\sigma_R = \sqrt{\frac{n_1 \cdot n_2(n_1 + n_2 + 1)}{12}}$$

$$= \sqrt{\frac{12 \cdot 12(12 + 12 + 1)}{12}} = 17.321$$

$$Z = \frac{R - \mu_R}{\sigma_R} = \frac{141 - 150}{17.321} = -0.520$$

$-1.96 \quad \uparrow \ 0 \quad 1.96$
-0.520

9. continued
Do not reject the null hypothesis. There is not enough evidence to support the claim that there is a difference in the amount of money awarded to the cities.

11.
H_0: There is no difference in the times needed to assemble the product.
H_1: There is a difference in the times needed to assemble the product. (claim)
C. V. $= \pm 1.96$

1.6	1.7	1.9	2.0	2.4	2.6	2.7	2.9	3.0
1	2	3	4	5	6	7	8	9
N	N	N	N	N	N	N	N	G

3.1	3.2	3.4	3.6	3.8	3.9	4.2	4.4	4.7
10	11	12	13	14	15	16	17	18
N	G	N	G	N	N	G	G	G

5.3	5.6	5.8	6.3	6.4	7.1	7.3	8.2
19	20	21	22	23	24	25	26
N	G	G	G	G	G	G	G

R = 245   (for graduates)

$$\mu_R = \frac{n_1(n_1 + n_2 + 1)}{2} = \frac{13(13 + 13 + 1)}{2}$$

$$= \frac{13(27)}{2} = 175.5$$

$$\sigma_R = \sqrt{\frac{n_1 \cdot n_2(n_1 + n_2 + 1)}{12}}$$

$$= \sqrt{\frac{13 \cdot 13(13 + 13 + 1)}{12}} = 19.5$$

$$Z = \frac{R - \mu_R}{\sigma_R} = \frac{245 - 175.5}{19.5} = \frac{69.5}{19.5} = 3.56$$

$-1.96 \quad 0 \quad 1.96 \quad \uparrow$
3.56

Reject the null hypothesis. There is enough evidence to support the claim that there is a difference in the productivity of the two groups.

EXERCISE SET 13-5

1.

The t-test for dependent samples.

3.

B	A	B − A	|B − A|	Rank	Signed Rank
108	110	− 2	2	1	− 1
97	97	0			
115	103	12	12	4.5	4.5
162	168	− 6	6	2	− 2
156	143	13	13	6	6
105	112	− 7	7	3	− 3
153	141	12	12	4.5	4.5

Sum of the − ranks: $(-1)+(-2)+$
$(-3)=-6$.
Sum of the + ranks: $4.5+6+4.5=15$
$w_s = 6$

5.
C. V. = 20 $w_s = 18$
Since $18 \leq 20$, reject the null hypothesis.

7.
C. V. = 130 $w_s = 142$
Since $142 > 130$, do not reject the null hypothesis.

9.
H_0: The workshop did not reduce anxiety.
H_1: The workshop reduced anxiety. (claim)

B	A	B − A	|B − A|	Rank	Signed Rank
23	22	1	1	1.5	1.5
26	29	− 3	3	6	− 6
30	27	3	3	6	6
31	29	2	2	3.5	3.5
39	33	6	6	8	8
23	21	2	2	3.5	3.5
28	25	3	3	6	6
27	28	− 1	1	1.5	− 1.5

Sum of the − ranks:
$(-6)+(-1.5)=-7.5$
Sum of the + ranks:
$1.5+6+3.5+8+3.5+6=28.5$

$n = 8$ C. V. = 6
$w_s = 7.5$
Since $7.5 > 6$, do not reject the null hypothesis. There is not enough evidence to support the claim that the workshop reduced the anxiety of the subjects.

11.
H_0: The sizes of police forces have decreased or remained the same.
H_1: The sizes of police forces have increased. (claim)

$n = 10$ $\alpha = 0.05$ C. V. = 11

B	A	B − A	|B − A|	Rank	Signed Rank
23,339	29,327	− 5988	5988	10	− 10
6886	7637	− 751	751	7	− 7
12,353	12,093	260	260	2	2
3716	4734	− 1018	1018	9	− 9
7218	6225	993	993	8	8
1376	1861	− 485	485	4	− 4
3808	3860	− 52	52	1	− 1
2084	2807	− 723	723	6	− 6
1635	1978	− 343	343	3	− 3
1159	1662	− 503	503	5	− 5

Sum of the − ranks:
$(-10)+(-7)+(-9)+(-4)+(-1)$
$+(-6)+(-3)+(-5)=-45$

Sum of the + ranks: $2+8=10$

$w_s = 10$

Since $10 \leq 11$, reject the null hypothesis. There is enough evidence to support the claim that the police forces have increased.

13.
H_0: There is no change in the size of the rosters. (claim)
H_1: The size of the rosters has changed.

$n = 8$ $\alpha = 0.10$ C. V. = 6

B	A	B − A	|B − A|	Rank	Signed Rank
33	33	0			
29	37	− 8	8	5.5	− 5.5
42	46	− 4	4	4	− 4
32	34	− 2	2	1	− 1
28	45	− 17	17	7	− 7
55	58	− 3	3	2.5	− 2.5
46	46	0			
26	23	3	3	2.5	2.5
17	35	− 18	18	8	− 8
30	22	8	8	5.5	5.5
45	45	0			
21	21	0			
42	42	0			

Sum of the + ranks: $2.5+5.5=8$

13. continued
Sum of the − ranks:

$$(-5.5) + (-4) + (-1) + (-7) +$$

$$(-2.5) + (-8) = -28$$

$$w_s = 8$$

Since $8 > 6$, do not reject the null hypothesis. There is not enough evidence to reject the claim that there is no change in the rosters.

EXERCISE SET 13-6

1.
H_0: There is no difference in the number of calories each brand contains.
H_1: There is a difference in the number of calories each brand contains. (claim)
C. V. $= 7.815$ $\alpha = 0.05$ d. f. $= 3$

A	Rank	B	Rank	C	Rank	D	Rank
112	7	110	6	109	5	106	3
120	13	118	12	116	9.5	122	15
135	24	123	16	125	17.5	130	21.5
125	17.5	128	19.5	130	21.5	117	11
108	4	102	2	128	19.5	116	9.5
121	14	101	1	132	23	114	8
$R_1=$	79.5	$R_2=$	56.5	$R_3=$	96	$R_4=$	68

$$H = \frac{12}{N(N+1)} \left(\frac{R_1^2}{n_1} + \frac{R_2^2}{n_2} + \frac{R_3^2}{n_3} + \frac{R_4^2}{n_4} \right) - 3(N+1)$$

$$H = \frac{12}{12(24+1)} \left(\frac{79.5^2}{6} + \frac{56.5^2}{6} + \frac{96^2}{6} \right.$$

$$\left. + \frac{68^2}{6} \right) = 2.842$$

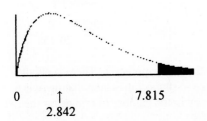

Do not reject the null hypothesis. There is not enough evidence to reject the claim that there is no difference in the calories.

3.
H_0: There is no difference in the sales of the stores.
H_1: There is a difference in the sales of the stores. (claim)

3. continued
C. V. $= 4.605$ d. f. $= 2$ $\alpha = 0.10$

Radio	Rank	TV	Rank	Paper	Rank
832	13	1024	18	329	2.5
648	11	996	16	437	5
562	10	1011	17	561	9
786	12	853	14	329	2.5
452	6	471	7	382	4
975	15	$R_2=$	72	495	8
$R_1=$	67			262	1
				$R_3=$	32

$$H = \frac{12}{N(N+1)} \left(\frac{R_1^2}{n_1} + \frac{R_2^2}{n_2} + \frac{R_3^2}{n_3} + \frac{R_4^2}{n_4} \right)$$

$$- 3(N+1)$$

$$= \frac{12}{18(18+1)} \left(\frac{67^2}{6} + \frac{72^2}{5} + \frac{32^2}{7} \right) - 3(18+1)$$

$$= \frac{12}{342} (748.167 + 1036.8 + 146.286) - 3(19)$$

$$H = 10.8$$

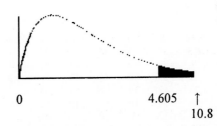

Reject the null hypothesis. There is enough evidence to support the claim that there is a difference in sales.

5.
H_0: There is no difference in the yields of the three plots.
H_1: There is a difference in the yields of the three plots. (claim)
C. V. $= 9.210$ d. f. $= 2$ $\alpha = 0.01$

A	Rank	B	Rank	C	Rank
32	2	43	6	50	10
38	3	45	7	56	14
31	1	49	9	58	15
40	5	46	8	54	13
39	4	51	11	52	12
$R_1=$	15	$R_2=$	41	$R_3=$	64

$$H = \frac{12}{N(N+1)} \left(\frac{R_1^2}{n_1} + \frac{R_2^2}{n_2} + \frac{R_3^2}{n_3} + \frac{R_4^2}{n_4} \right) - 3(N+1)$$

$$= \frac{12}{15(15+1)} \left(\frac{15^2}{5} + \frac{41^2}{5} + \frac{64^2}{5} \right) - 3(15+1)$$

5. continued

H = 12.020

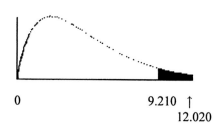

0   9.210 ↑
                         12.020

Reject the null hypothesis. There is enough evidence to support the claim that the yields of the three plots are different.

7.

H_0: There is no difference in the number of deaths due to lightning, tornado, flood or blizzard.

H_1: There is a difference in the number of weather-related deaths. (claim)

C. V. = 6.251

L	Rank	T	Rank	F	Rank	B	Rank
39	6.5	30	1.5	46	12	54	16
41	9	39	6.5	55	17	43	10
73	22	39	6.5	45	11	39	6.5
74	23	53	15	109	24	35	4
67	20	50	14	62	19	56	18
68	21	32	3	30	1.5	48	13
R_1=	101.5	R_2=	46.5	R_3=	85.5	R_4=	67.5

$$H = \frac{12}{N(N+1)} \left(\frac{R_1^2}{n_1} + \frac{R_2^2}{n_2} + \frac{R_3^2}{n_3} + \frac{R_4^2}{n_4} \right) - 3(N+1)$$

$$H = \frac{12}{24(24+1)} \left(\frac{101.5^2}{6} + \frac{46.5^2}{6} + \frac{85.5^2}{6} + \right.$$

$$\left. \frac{67.5^2}{6} \right) - 3(24+1)$$

H = 5.537

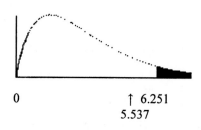

0   ↑ 6.251
                    5.537

Do not reject the null hypothesis. There is not enough evidence to support the claim that there is a difference in the number of weather-related deaths.

9.

H_0: There is no difference in the number of crimes in the 5 precincts.

H_1: There is a difference in the number of crimes in the 5 precincts. (claim)

C. V. = 13.277   d. f. = 4 $\alpha = 0.01$

1	Rank	2	Rank	3	Rank
105	24	87	13	74	7.5
108	25	86	12	83	11
99	22	91	16	78	9
97	20	93	18	74	7.5
92	17	82	10	60	5
R_1=	108	R_2=	69	R_3=	40

4	Rank	5	Rank
56	3	103	23
43	1	98	21
52	2	94	19
58	4	89	15
62	6	88	14
R_4=	16	R_5=	92

$$H = \frac{12}{N(N+1)} \left(\frac{R_1^2}{n_1} + \frac{R_2^2}{n_2} + \frac{R_3^2}{n_3} + \frac{R_4^2}{n_4} + \frac{R_5^2}{n_5} \right) - 3(N+1)$$

$$= \frac{12}{25(25+1)} \left(\frac{108^2}{5} + \frac{69^2}{5} + \frac{40^2}{5} + \frac{16^2}{5} \right.$$

$$\left. + \frac{92^2}{5} \right) - 3(25+1) = 20.753$$

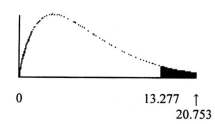

0   13.277 ↑
                         20.753

Reject the null hypothesis. There is enough evidence to support the claim that there is a difference in the number of crimes for the precincts.

11.

H_0: There is no difference in the final exam scores of the three groups.

H_1: There is a difference in the final exam scores of the three groups. (claim)

d. f. = 2 $\alpha = 0.10$

11. continued

A	Rank	B	Rank	C	Rank
98	18	97	17	99	19
100	20	88	11	94	14
95	15	82	8	96	16
92	13	84	9	89	12
86	10	75	5	81	7
76	6	73	3	72	2
71	1	74	4	$R_3 =$	70
$R_1 =$	83	$R_2 =$	57		

$$H = \frac{12}{N(N+1)} \left(\frac{R_1^2}{n_1} + \frac{R_2^2}{n_2} + \frac{R_3^2}{n_3} \right) - 3(N+1)$$

$$= \frac{12}{20(20+1)} \left(\frac{83^2}{7} + \frac{57^2}{7} + \frac{70^2}{6} \right) - 3(20+1)$$

H = 1.710

P-value > 0.10 (0.425)
Do not reject. There is not enough evidence to support the claim that there is a difference in the final exam scores of the three groups.

EXERCISE SET 13-7

1.
0.716

3.
0.648

5.
H_0: $\rho = 0$
H_1: $\rho \neq 0$
C. V. $= \pm 0.564$ $n = 10$ $\alpha = 0.10$

Tornadoes	R_1	Temp	R_2	$R_1 - R_2$	d^2
668	6	112	5	1	1
781	7	118	9	-2	4
1590	10	109	3	7	49
798	8	117	8	0	0
1198	9	121	10	-1	1
169	2	108	2	0	0
310	3	111	4	-1	1
360	4	113	6	-2	4
21	1	105	1	0	0
625	5	114	7	-2	4
				$\sum d^2 =$	64

$$r_s = 1 - \frac{6 \cdot \sum d^2}{n(n^2 - 1)} = 1 - \frac{6 \cdot 64}{10(10^2 - 1)}$$

$r_s = 0.612$

Reject the null hypothesis. There is a significant relationship between the number of tornadoes and high temperatures.

7.
H_0: $\rho = 0$
H_1: $\rho \neq 0$

C. V. $= \pm 0.886$ $n = 6$ $\alpha = 0.05$

Sentence	R_1	Time	R_2	$R_1 - R_2$	d^2
227	6	97	6	0	0
120	5	45	5	0	0
106	4	40	4	0	0
77	3	26	3	0	0
60	2	18	1	1	1
49	1	21	2	-1	1
				$\sum d^2 =$	2

$$r_s = 1 - \frac{6 \cdot \sum d^2}{n(n^2 - 1)}$$

$$r_s = 1 - \frac{6 \cdot 2}{6(6^2 - 1)} = 0.943$$

Reject. There is a significant relationship between sentence and time served.

9.
H_0: $\rho = 0$
H_1: $\rho \neq 0$

C. V. $= \pm 0.738$

Teen	R_1	Parent	R_2	$R_1 - R_2$	d^2
4	4	1	1	3	9
6	6	7	7	-1	1
2	2	5	5	-3	9
8	8	4	4	4	16
1	1	3	3	-2	4
7	7	8	8	-1	1
3	3	2	2	1	1
5	5	6	6	-1	3
				$\sum d^2 =$	42

$$r_s = 1 - \frac{6 \cdot 42}{8(8^2 - 1)} = 1 - \frac{252}{504} = 0.5$$

Do not reject the null hypothesis. There is not enought evidence to say that a significant relationship exists between the rankings.

11.
H_0: $\rho = 0$
H_1: $\rho \neq 0$
C. V. $= \pm 0.886$

11. continued

Designer	R_1	Customer	R_2	$R_1 - R_2$	d^2
48	2	35	2	0	0
76	4	44	3	1	1
30	1	28	1	0	0
88	5	50	4	1	1
61	3	75	5	-2	4
93	6	85	6	0	0
				$\sum d^2 =$	6

$$r_s = 1 - \frac{6\sum d^2}{n(n^2-1)} = 1 - \frac{6 \cdot 6}{6(6^2-1)}$$

$$r_s = 0.829$$

Do not reject the null hypothesis. There is not enough evidence to say that a significant relationship exists between the rankings.

13.
H_0: $\rho = 0$
H_1: $\rho \neq 0$

C. V. $= \pm 0.591$

Engineer	R_1	Customer	R_2	$R_1 - R_2$	d^2
81	8	85	9	-1	1
70	7	75	7	0	0
65	6	68	6	0	0
54	5	50	4	1	1
43	4	52	5	-1	1
90	12	95	11	1	1
41	3	48	3	0	0
88	11	100	12	-1	1
40	2	44	2	0	0
85	10	90	10	0	0
82	9	83	8	1	1
35	1	20	1	0	0
				$\sum d^2 =$	6

$$r_s = 1 - \frac{6\sum d^2}{n(n^2-1)} = 1 - \frac{6 \cdot 6}{12(12^2-1)}$$

$$r_s = 0.979$$

Reject the null hypothesis. There is enough evidence to say that a significant relationship exists between the rankings.

15.
H_0 = The occurrance of cavities is random.
H_1 = The null hypothesis is not true.

The median of the data set is two. Using A = above and B = below the runs (going across) are shown:

B AA B AAA B A BB AAAA B A B A B A
B A B AAA B A BB

15. continued
There are 21 runs. The expected number of runs is between 10 and 22; therefore, the null hypothesis should not be rejected. The number of cavities occurs at random.

17.
H_0: The lotto numbers occur at random.
H_1: The null hypothesis is not true.

OO E OO EE O EE OOO EE OO E O E O
EE

There are 14 runs and this is between 7 and 18; hence, do not reject the null hypothesis. The numbers occur at random.

19.
H_0: The number of defective cigarettes manufactured by a machine occurs at random.
H_1: The null hypothesis is not true.

D AAAAAA DD A DD AAA DD AAAAAAA
DDD AAA

There are 10 runs and since this is between 9 and 20, the null hypothesis is not rejected. The defective cigarettes occur at random.

21.
H_0: The number of absences of employees occur at random.
H_1: The null hypothesis is not true.

The median of the data is 12. Using A = above and B = below, the runs are shown.

A B AAAAAAA BBBBBBBBB AAAAAA
BBBB
There are six runs. The expected number of runs is between 9 and 21, hence the null hypothesis is rejected since six is not between 9 and 21. The number of absences do not occur at random.

23.
H_0: The I. Q.'s are random.
H_1: The null hypothesis is not true.

The median of the data is 98. Using A = above and B = below, the runs are shown.

AAAAAAA BBBBB A BB A BB AA
Since there are 7 runs, the null hypothesis is not rejected because the 7 is within the 6 to 16 range. The I. Q.'s are random.

25.

$$r = \frac{\pm z}{\sqrt{n-1}} = \frac{\pm 2.58}{\sqrt{30-1}} = \pm 0.479$$

27.

$$r = \frac{\pm z}{\sqrt{n-1}} = \frac{\pm 1.65}{\sqrt{60-1}} = \pm 0.215$$

REVIEW EXERCISES - CHAPTER 13

1.

$$+ + + + - - - - - - + +$$

H_0: median $= 5$ (claim)
H_1: median $\neq 5$

C. V. $= 2$ Test Value $= 6$

Do not reject. There is not enough evidence to reject the claim that the median is 5.

3.

$$- + - - - - - + +$$

H_0: The special diet has no effect on weight.
H_1: The diet increases weight. (claim)

C. V. $= 1$ Test value $= 3$

Do not reject the null hypothesis. There is not enough evidence to support the claim that there was an increase in weight.

5.

H_0: There is no difference in the amounts of money each group spent for the textbook.
H_1: There is a difference in the amounts of money each group spent for the textbook. (claim)
C. V. $= \pm 1.65$

36	46	48	49	50	51	52	53	55
1	2	3	4	5	6	7	8	9
B	B	B	B	B	B	B	B	E

58	58	62	63	63	64	72	73
10.5	10.5	12	13.5	13.5	15	16	17
B	E	B	B	E	E	E	E

74	78	78	85	86	88	93	98
18	19.5	19.5	21	22	23	24	25
B	E	E	E	E	E	E	E

$R = 90$

$$\mu_R = \frac{n_1(n_1 + n_2 + 1)}{2} = \frac{12(12 + 13 + 1)}{2} = 156$$

5. continued

$$\sigma_R = \sqrt{\frac{n_1 n_2(n_1 + n_2 + 1)}{12}}$$

$$\sigma_R = \sqrt{\frac{12 \cdot 13(12 + 13 + 1)}{12}} = 18.38$$

$$Z = \frac{R - \mu_R}{\sigma_R} = \frac{90 - 156}{18.38} = \frac{-66}{18.38} = -3.59$$

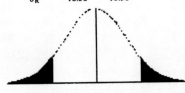

$$\uparrow \quad -1.65 \qquad 0 \qquad 1.65$$
$$-3.59$$

Reject the null hypothesis. There is enough evidence to support the claim that there is a difference in the amount spent on the textbooks.

7.

H_0: The number of sick days workers used was not reduced.
H_1: The number of sick days workers used was reduced. (claim)

B	A	B – A	|B – A|	Rank	Signed Rank
6	8	– 2	2	1.5	– 1.5
15	12	3	3	3.5	3.5
18	16	2	2	1.5	1.5
14	9	5	5	6	6
27	23	4	4	5	4
17	14	3	3	3.5	3
9	15	– 6	6	7	– 7

Sum of the + ranks: 18
Sum of the – ranks: – 8.5
$w_s = 8.5$
C. V. $= 4$ $\alpha = 0.05$ n $= 7$

Do not reject the null hypothesis. There is not enough evidence to support the claim that the number of sick days was reduced.

9.

H_0: The diet has no effect on learning.
H_1: The diet affects learning. (claim)

C. V. $= 5.991$ d. f. $= 2$ $\alpha = 0.05$

9. continued

Diet 1	Rank	Diet 2	Rank	Diet 3	Rank
8	11	2	1	9	13.5
6	6.5	3	2	15	17.5
12	15	6	6.5	17	19
15	17.5	8	11	8	11
9	13.5	7	8.5	4	3.5
7	8.5	4	3.5	13	16
5	5	$R_2 =$	32.5	18	20
$R_1 =$	77			20	21
				$R_3 =$	121.5

$$H = \frac{12}{N(N+1)} \left(\frac{R_1^2}{n_1} + \frac{R_2^2}{n_2} + \frac{R_3^2}{n_3} \right) - 3(N+1)$$

$$= \frac{12}{21(21+1)} \left(\frac{77^2}{7} + \frac{32.5^2}{6} + \frac{121.5^2}{8} \right)$$

$$- 3(21 + 1) = 8.5$$

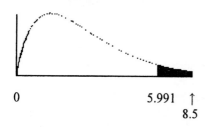

0   5.991 ↑
                        8.5

Reject the null hypothesis. There is enough evidence to support the claim that the diets do affect learning.

11.

Boys	R_1	Girls	R_2	$R_1 - R_2$	d^2
3	3	4	4	−1	1
2	2	5	5	−3	9
6	6	1	1	5	25
1	1	3	3	−2	4
5	5	2	2	3	9
4	4	6	6	−2	4
				$\sum d^2 =$	52

$$r_s = 1 - \frac{6 \sum d^2}{6(6^2 - 1)} = 1 - \frac{6(52)}{6(35)} = -0.486$$

H_0: $\rho = 0$
H_1: $\rho \neq 0$

C. V. $= \pm 0.886$

Do not reject the null hypothesis. There is not enough evidence to say that a relationship exists between the rankings of the boys and girls.

13.
H_0: The grades of the students who finish the exam occur at random.
H_1: The null hypothesis is not true.

The median grade is 73. Using A = above and B = below, the runs are:

AAAA B AAAA BBBB AAAAA BB A BBBBBBB

Since there are eight runs and this does not fall between 9 and 21, the null hypothesis is rejected. The grades do not occur at random.

CHAPTER 13 QUIZ

1. False
2. False, they are less sensitive.
3. True
4. True
5. a.
6. c.
7. d.
8. b.
9. non-parametric
10. nominal, ordinal
11. sign
12. sensitive

13. H_0: Median = 300 (claim)
H_1: Median ≠ 300
There are seven + signs. Do not reject since 7 is greater than the critical value 5. There is not enough evidence to reject the claim that the median is 300.

14. H_0: Median = 1200 (claim)
H_1: Median ≠ 1200
There are ten − signs. Do not reject since 10 is greater than the critical value 6. There is not enough evidence to reject the claim that the median is 1,200.

15. H_0: There will be no change in the weight of the turkeys after the special diet.
H_1: The turkeys will weigh more after the special diet. (claim)
There is one + sign. Reject the null hypothesis. There is enough evidence to support the claim that the turkeys gained weight on the special diet.

16. H_0: The distributions are the same.
H_1: The distributions are different. (claim)
C. V. $= \pm 1.96$
z $= -0.05$

16. continued
Do not reject the null hypothesis. There is not enough evidence to reject the claim that the distributions are the same.

17. H_0: The distributions are the same.
H_1: The distributions are different. (claim)
C. V. $= \pm 1.65$
$z = -0.14$
Do not reject the null hypothesis. There is not enough evidence to say there is a difference in the costs of the textbooks.

18. H_0: There is no difference in the GPA's before and after the workshop.
H_1: There is a difference in GPA's before and after the workshop. (claim)
C. .V. $= 2$ Test statistic $= 0$
Reject the null hypothesis. There is enough evidence to support the claim that there is a difference in the GPA's of the students.

19. H_0: There is no difference in the breaking strengths of the tapes.
H_1: There is a difference in breaking strengths. (claim)
$H = 29.25$
$\chi^2 = 5.991$
Reject the null hypothesis. There is enough evidence to support the claim that there is a difference in the breaking strengths of the tapes.

20. H_0: There is no difference in reaction times.
H_1: There is a difference in reaction times. (claim)
$H = 6.9$
$0.025 < $ P-value $ < 0.05$ (0.032)
Reject the null hypothesis. There is enough evidence to support the claim that there is a difference in the reaction times of the monkeys.

21. H_0: $\rho = 0$
H_1: $\rho \neq 0$
C. V. $= \pm 0.700$
$r = 0.846$
Reject the null hypothesis. There is a significant relationship between the number of homework exercises and exam scores.

22. H_0: $\rho = 0$
H_1: $\rho \neq 0$
$r = -0.400$
C. V. $= 0.900$
Do not reject the null hypothesis. There is not enough evidence ot say that a significant relationship

22. continued
exists between the rankings of the brands by males and females.

23. H_0: The gender of babies occurs at random.
H_1: The null hypothesis is false.
$\alpha = 0.05$ C. V. $= 8, 19$
There are 10 runs, which is between 8 and 19. Do not reject the null hypothesis. There is not enough evidence to reject the null hypothesis that the gender occurs at random.

24. H_0: There is no difference in output ratings.
H_1: There is a difference in output ratings after reconditioning. (claim)
$\alpha = 0.05$ $n = 9$ C. V. $= 6$
test statistic $= 0$
Do not reject the null hypothesis. There is not enough evidence to support the claim that there is a difference in output ratings before and after reconditioning.

25. H_0: The numbers occur at random
H_1: The null hypothesis is false.
$\alpha = 0.05$ C. V. $= 9, 21$
The median number is 538.
There are 16 runs, reading from left to right, and since this is between 9 and 21, the null hypothesis is not rejected. There is not enough evidence to reject the null hypothesis that the numbers occur at random.

EXERCISE SET 14-2

1.
Random, systematic, stratified, cluster.

3.
A sample must be randomly selected.

5.
Talking to people on the street, calling people on the phone, and asking one's friends are three incorrect ways of obtaining a sample.

7.
Random sampling has the advantage that each unit of the population has an equal chance of being selected. One disadvantage is that the units of the population must be numbered, and if the population is large this could be somewhat time consuming.

9.
An advantage of stratified sampling is that it ensures representation for the groups used in stratification; however, it is virtually impossible to stratify the population so that all groups could be represented.

11 through 19.
Answers will vary.

EXERCISE SET 14-3

1.
This is a biased question. Change the question to read: "Do you think XYZ Department Store should carry brand-name merchandise?"

3. This is a biased question. Change the question to read: "Should banks charge a fee to balance their customers' checkbooks?"

5.
This question has confusing wording. Change the question to read: "How many hours did you study for this exam?"

7.
This question has confusing wording. Change the question to read: "If a plane were to crash on the border of New York and New Jersey, where should the victims be buried?"

9.
Answers will vary.

EXERCISE SET 14-5

1.
Simulation involves setting up probability experiments that mimic the behavior of real life events.

3.
John Van Neumann and Stanislaw Ulam.

5.
The steps are:
  1. List all possible outcomes.
  2. Determine the probability of each outcome.
  3. Set up a correspondence between the outcomes and the random numbers.
  4. Conduct the experiment using random numbers.
  5. Repeat the experiment and tally the outcomes.
  6. Compute any statistics and state the conclusions.

7.
When the repetitions increase there is a higher probability that the simulation will yield more precise answers.

9.
Use random numbers one through eight to make a shot and nine and zero to represent a "miss".

11.
Use random numbers one through seven to represent a "hit" and eight, nine and zero to represent a "miss".

13.
Let an odd number represent "heads" and an even number to represent "tails"; then each person selects a digit at random.

15 through 23.
Answers will vary.

REVIEW EXERCISES - CHAPTER 15

1 - 7.
Answers will vary.

9.
Use two digit random numbers 01 through 65 to represent a strike-out and 66 through 99 and 00 to represent anything other than a strike-out.

11.
Select two digits between one and six to represent the dice.

13.
Let the digits 1 – 3 represent "rock"
Let the digits 4 – 6 represent "paper"
Let the digits 7 – 9 represent "scissors"
Omit 0.

15.
Answers will vary.

17.
Answers will vary.

19.
This is a biased question. Change the question to read: "Have you ever driven through a red light?"

21.
This is a double-barreled question. Change the question to read: "Do you think all automobiles should have heavy-duty bumpers?"

CHAPTER 14 QUIZ

1. True
2. True
3. False, only random numbers generated by a random number table are random.
4. True
5. a.
6. c.
7. c.
8. larger
9. biased
10. cluster
11. Answers will vary.
12. Answers will vary.
13. Answers will vary.
14. Answers will vary.
15. Use two-digit random numbers. 01 through 35 constitute a win and 36 through 00 constitute a loss.
16. Use two-digit random numbers. 01 - 03 constitute a cancellation.

17. Use two-digit random numbers. 01 - 13 for cards.
18. Use random numbers 1 - 6 to simulate the roll of a die and random numbers 01 - 13 to simulate the cards.
19. Use random numbers 1 - 6 to simulate a roll of a die.
20 - 24. Answers will vary.

A-1

A-1. $9! = 9 \cdot 8 \cdot 7 \cdot 6 \cdot 5 \cdot 4 \cdot 3 \cdot 2 \cdot 1 = 362,880$

A-3. $5! = 5 \cdot 4 \cdot 3 \cdot 2 \cdot 1 = 120$

A-5. $1! = 1$

A-7. $\frac{12!}{9!} = \frac{12 \cdot 11 \cdot 10 \cdot 9!}{9!} = 1320$

A-9. $\frac{5!}{3!} = \frac{5 \cdot 4 \cdot 3!}{3!} = 20$

A-11. $\frac{9!}{(4!)(5!)} = \frac{9 \cdot 8 \cdot 7 \cdot 6 \cdot 5!}{4 \cdot 3 \cdot 2 \cdot 1 \cdot 5!} = 126$

A-13. $\frac{8!}{4!4!} = \frac{8 \cdot 7 \cdot 6 \cdot 5 \cdot 4!}{4 \cdot 3 \cdot 2 \cdot 1 \cdot 4!} = 70$

A-15. $\frac{10!}{(10!)(0!)} = \frac{10!}{10! \cdot 1} = 1$

A-17. $\frac{8!}{3!3!2!} = \frac{8 \cdot 7 \cdot 6 \cdot 5 \cdot 4 \cdot 3!}{3! \cdot 3 \cdot 2 \cdot 1 \cdot 2 \cdot 1} = 560$

A-19. $\frac{10!}{3!2!5!} = \frac{10 \cdot 9 \cdot 8 \cdot 7 \cdot 6 \cdot 5!}{3 \cdot 2 \cdot 1 \cdot 2 \cdot 1 \cdot 5!} = 2520$

A-2

A-21.

X	X^2	$X - \overline{X}$	$(X - \overline{X})^2$
9	81	− 3.1	9.61
17	289	4.9	24.01
32	1024	19.9	396.01
16	256	3.9	15.21
8	64	− 4.1	16.81
2	4	− 10.1	102.01
9	81	− 3.1	9.61
7	49	− 5.1	26.01
3	9	− 9.1	82.81
18	324	5.9	34.81
121	2181		716.9

$\sum X = 121$ $\quad \overline{X} = \frac{121}{10} = 12.1$

$\sum X^2 = 2,181$ $\quad (\sum X)^2 = 121^2 = 14,641$

$\sum (X - \overline{X})^2 = 716.9$

A-23.

X	X^2	$X - \overline{X}$	$(X - \overline{X})^2$
5	25	− 1.4	1.96
12	144	5.6	31.36
8	64	1.6	2.56
3	9	− 3.4	11.56
4	16	− 2.4	5.76
32	258		53.20

$\sum X = 32$ $\quad \overline{X} = \frac{32}{5} = 6.4$

$\sum X^2 = 258$ $\quad (\sum X)^2 = 32^2 = 1,024$

$\sum (X - \overline{X})^2 = 53.2$

A-25.

X	X^2	$X - \overline{X}$	$(X - \overline{X})^2$
80	6400	14.4	207.36
76	5776	10.4	108.16
42	1764	− 23.6	556.96
53	2809	− 12.6	158.76
77	5929	11.4	129.96
328	22678		1161.20

$\sum X = 328$ $\quad \overline{X} = \frac{328}{5} = 65.6$

$\sum X^2 = 22,678$ $\quad (\sum X)^2 = 328^2 = 107,584$

$\sum (X - \overline{X})^2 = 1161.2$

A-27.

X	X^2	$X - \overline{X}$	$(X - \overline{X})^2$
53	2809	− 16.3	265.69
72	5184	2.7	7.29
81	6561	11.7	136.89
42	1764	− 27.3	745.29
63	3969	− 6.3	39.69
71	5041	1.7	2.89
73	5329	3.7	13.69
85	7225	15.7	246.49
98	9604	28.7	823.69
55	3025	− 14.3	204.49
693	50511		2486.10

$\sum X = 693$ $\quad \overline{X} = \frac{693}{10} = 69.3$

$\sum X^2 = 50,511$ $\quad (\sum X)^2 = 693^2 = 480,249$

$\sum (X - \overline{X})^2 = 2486.1$

A-29.

X	X²	X − X̄	(X − X̄)²
12	144	− 41	1681
52	2704	− 1	1
36	1296	− 17	289
81	6561	28	784
63	3969	10	100
74	5476	21	441
318	20150		3296

$$\sum X = 318 \qquad \bar{X} = \frac{318}{6} = 53$$

$$\sum X^2 = 20{,}150 \qquad (\sum X)^2 = 318^2 = 101{,}124$$

$$\sum (X - \bar{X})^2 = 3{,}296$$

A-3

A-31.

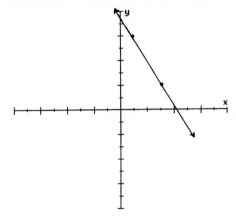

A-33.

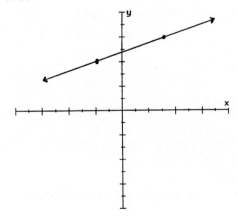

A-35.

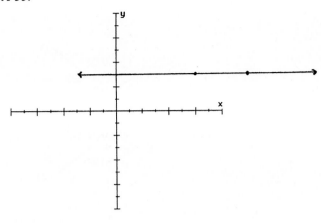

A-37.

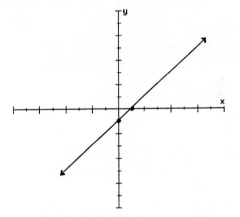

Two points are: (1, 0) and (0, -1).

A-39.

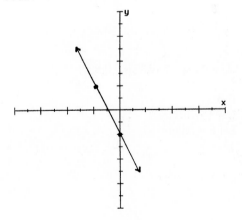

Two points are: (-2, 2) and (0, -2).

B-2

B-1.

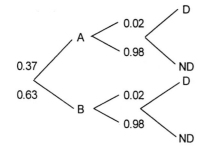

$$P(B \mid \text{defective}) = \frac{P(B) \cdot P(\text{def} \mid B)}{P(A) \cdot P(\text{def} \mid A) + P(B) \cdot P(\text{def} \mid B)}$$

$$= \frac{0.63(.02)}{0.37(0.02) + 0.63(0.02)} = \frac{0.0126}{0.02} = 0.63$$

B-3.
Let D = person has disease
Let $\overline{D}$ = person does not have disease
Let A = positive test result
Let $\overline{A}$ = negative test result

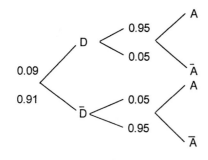

$$P(D \mid A) = \frac{P(D) \cdot P(A \mid D)}{P(D) \cdot P(A \mid D) + P(\overline{D}) \cdot P(A \mid \overline{D})}$$

$$= \frac{0.09(0.95)}{0.09(0.95) + 0.91(0.05)} = 0.653$$

B-5.
Let S = success
Let F = failure

B-5 continued

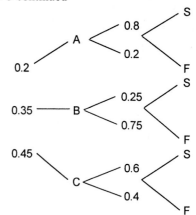

$$P(B \mid S) = \frac{P(B) \cdot P(S \mid B)}{P(A) \cdot P(S \mid A) + P(B) \cdot P(S \mid B) + P(C) \cdot P(S \mid C)}$$

$$= \frac{0.35(0.75)}{0.20(0.80) + 0.35(0.75) + 0.45(0.60)} = \frac{0.2625}{0.6925}$$

$$= 0.379$$

B-7.

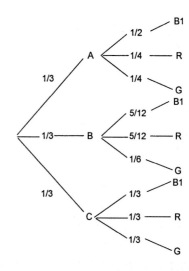

$$P(A \mid R) = \frac{P(A) \cdot P(R \mid A)}{P(A) \cdot P(R \mid A) + P(B) \cdot P(R \mid B) + P(C) \cdot P(R \mid C)}$$

$$= \frac{\frac{1}{3} \cdot \frac{1}{4}}{\frac{1}{3} \cdot \frac{1}{4} + \frac{1}{3} \cdot \frac{5}{12} + \frac{1}{3} \cdot \frac{1}{3}} = \frac{\frac{1}{12}}{\frac{1}{12} + \frac{5}{36} + \frac{1}{9}} = \frac{\frac{1}{12}}{\frac{1}{3}} = \frac{1}{4}$$

B-9.
Let J = traffic jam and $\overline{J}$ = no traffic jam

B-9 continued

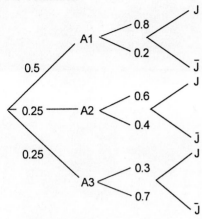

$P(A_1 | J) =$

$$\frac{P(A_1) \cdot P(J \mid A_1)}{P(A_1) \cdot P(J \mid A_1) + P(A_2) \cdot P(J \mid A_2) + P(A_3) \cdot P(J \mid A_3)}$$

$$\frac{0.5(0.8)}{0.5(0.8) + 0.25(0.6) + 0.25(0.3)} = \frac{0.4}{0.4 + 0.15 + 0.075}$$

$$= \frac{0.4}{0.625} = 0.64$$

B-11.

$P(A) = \frac{350}{1400} = 0.25 \quad P(B) = \frac{1050}{1400} = 0.75$

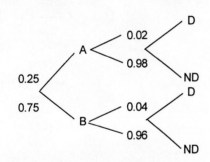

$P(B \mid D) = \frac{P(B) \cdot P(D \mid B)}{P(A) \cdot P(D \mid A) + P(B) \cdot P(D \mid B)}$

$$= \frac{0.75(0.04)}{0.25(0.02) + 0.75(0.04)} = \frac{0.03}{0.035} = 0.857$$

Notes

Notes

Notes

Notes

Notes

Notes

Notes

Notes